SpringerBriefs in Water Science and Technology

SpringerBriefs in Water Science and Technology present concise summaries of cutting-edge research and practical applications. The series focuses on interdisciplinary research bridging between science, engineering applications and management aspects of water. Featuring compact volumes of 50 to 125 pages (approx. 20,000–70,000 words), the series covers a wide range of content from professional to academic such as:

- Literature reviews
- In-depth case studies
- Bridges between new research results
- Snapshots of hot and/or emerging topics

Topics covered are for example the movement, distribution and quality of freshwater; water resources; the quality and pollution of water and its influence on health; and the water industry including drinking water, wastewater, and desalination services and technologies.

Both solicited and unsolicited manuscripts are considered for publication in this series.

Manish Kumar • Durga Prasad Panday

Enabling Water Security and Sustainability: Global Challenges and Solutions

Integrated Water Resources Management

Springer

Manish Kumar
Department of Sustainable Technology and Civil Engineering
School of Engineering and Sciences
Monterrey, Nuevo León, Mexico

School of Advanced Engineering, UPES
Sustainability Cluster
Dehradun, Uttarakhand, India

Durga Prasad Panday
Department of Sustainable Technology and Civil Engineering
School of Engineering and Sciences
Monterrey, Nuevo León, Mexico

School of Advanced Engineering, UPES
Sustainability Cluster
Dehradun, Uttarakhand, India

ISSN 2194-7244 ISSN 2194-7252 (electronic)
SpringerBriefs in Water Science and Technology
ISBN 978-3-032-21520-8 ISBN 978-3-032-21521-5 (eBook)
https://doi.org/10.1007/978-3-032-21521-5

This Springer imprint is published by the registered company Springer Nature Switzerland AG
The registered company address is: Gewerbestrasse 11, 6330 Cham, Switzerland

Foreword

Water security and sustainability are among the most pressing challenges of the twenty-first century. Climate change, rapid urbanization, groundwater depletion, emerging contaminants, and increasing resource demands are placing unprecedented stress on global water systems. Addressing these challenges requires interdisciplinary research, technological innovation, and sustainability-oriented policy frameworks. In this context, *Enabling Water Security and Sustainability: Global Challenges and Solutions* by Dr. Manish Kumar and Dr. Durga Prasad Panday presents a timely and significant contribution to the field of water science and sustainable development.

At UPES, sustainability, environmental resilience, and climate-responsive research form an important part of our academic and institutional vision. Through interdisciplinary initiatives in sustainability, environmental engineering, water resource management, ESG, and climate adaptation, the university actively promotes solutions aimed at addressing emerging global challenges. The themes explored in this book strongly align with these efforts.

This book is particularly valuable because of its comprehensive and well-structured approach. Beginning with a global overview of water quantity and quality distribution, the authors establish a strong foundation for understanding contemporary water challenges. The discussions on transboundary water management and sharing are especially relevant in today's world, where cooperation and integrated governance are essential for sustainable resource management.

I also appreciate the book's emphasis on innovative conservation practices, such as managed aquifer recharge (MAR), along with its detailed treatment of geogenic and emerging contaminants. These chapters effectively combine scientific understanding with practical relevance and highlight the growing need for predictive, technology-driven, and sustainability-focused water management strategies. The concluding discussion on the One Health framework further strengthens this book by recognizing the interconnectedness of water, environmental systems, ecosystems, and human well-being.

The authors deserve sincere congratulations for bringing together such an interdisciplinary and forward-looking book. Their efforts in connecting hydrology, sustainability, climate science, governance, and environmental health within a single framework are highly commendable. I am confident that this book will serve as a valuable resource for researchers, students, policymakers, industry professionals, and all stakeholders working toward a resilient and sustainable water future.

UPES
Misraspatti, Uttarakhand, India

Ram Sharma

Preface

Water remains at the center of global sustainability, human well-being, environmental resilience, and economic development. However, increasing pressures from climate change, urbanization, groundwater depletion, pollution, and unsustainable resource utilization are creating unprecedented challenges for water systems worldwide. These evolving challenges demand an integrated scientific understanding, interdisciplinary collaboration, and sustainable management strategies. In this context, *Enabling Water Security and Sustainability: Global Challenges and Solutions* was developed to provide a concise yet comprehensive perspective on some of the most important contemporary issues related to water security and sustainability.

This book explores multiple dimensions of water systems, beginning with the global distribution of water quantity and quality and progressing toward topics, such as transboundary water governance, managed aquifer recharge (MAR), geogenic contamination, emerging contaminants, groundwater sustainability, and the One Health framework. While preparing this book, our objective was to bridge scientific understanding with practical relevance and to highlight the interconnected nature of environmental, hydrological, and societal systems.

The development of this book would not have been possible without the support of institutions and individuals who continue to encourage interdisciplinary research and sustainability-focused innovation. We sincerely acknowledge Tecnológico de Monterrey, Mexico, for fostering an environment that supports global scientific collaboration and interdisciplinary sustainability research. We are equally grateful to UPES, India, for promoting research and academic initiatives in sustainability, environmental engineering, and climate resilience.

We also express our heartfelt appreciation to Scholars Sachin Tripathi, Shiwangi Dogra, Aanchal Kumari, and Kanika Dogra for their valuable assistance and support during the preparation of this book. Their dedication and contribution at different stages of this work are sincerely acknowledged.

Finally, we thank researchers, students, collaborators, and the wider scientific community whose work and discussions continue to inspire meaningful progress in water research and sustainability science. We hope this book will serve as a useful resource for researchers, students, policymakers, and professionals working toward resilient, sustainable, and future-ready water systems.

Monterrey, Nuevo León, Mexico — Manish Kumar
Dehradun, Uttarakhand, India — Durga Prasad Panday

Contents

Chapter 1
Global Overview of Water Quantity and Quality Distribution

Abstract Water is a fundamental resource underpinning human well-being, ecological balance, and sustainable development, yet its availability and quality are highly uneven worldwide. The present chapter provides a comprehensive examination of global freshwater resources, considering both surface water and groundwater within the broader hydrological cycle. It highlights the growing challenges of water scarcity and stress, arising from the widening gap between renewable supply and escalating demand across agricultural, industrial, and domestic sectors. Agriculture alone consumes most global withdrawals, while urbanisation and industrialisation are placing unprecedented pressure on already fragile systems. Beyond issues of quantity, the chapter explores dimensions of water quality, addressing both natural variability and intensifying anthropogenic impacts. Agricultural runoff, industrial discharges, and untreated urban wastewater continue to degrade water bodies. At the same time, emerging contaminants such as pharmaceuticals, endocrine disruptors, and microplastics pose new and poorly understood risks to human health and ecosystems. Regional disparities are examined, with a particular focus on water-scarce regions in Africa and Asia, and in small island states, alongside case studies from the water-abundant areas where contamination undermines security. The chapter further examines governance frameworks, policy instruments, and technological innovations to promote equitable and sustainable water management. These include integrated water resources management, transboundary cooperation, and new technologies such as desalination, wastewater recycling, and digital monitoring systems. Situating water within the United Nations Sustainable Development Goals, the discussion underscores its interlinkages with food security, public health, climate resilience, and ecosystem protection. Looking ahead, it emphasises the need for integrated solutions, nature-based approaches, and resilience-building strategies to secure freshwater resources for future generations.

Keywords Water resources · Global water distribution · Water scarcity · Water quality · Groundwater and surface water

M. Kumar, D. P. Panday, *Enabling Water Security and Sustainability: Global Challenges and Solutions*, SpringerBriefs in Water Science and Technology,
https://doi.org/10.1007/978-3-032-21521-5_1

1.1 Introduction

Water is one of the most essential natural resources for sustaining life, ecosystems, and human development. Despite covering about 71% of the Earth's surface, only a small fraction, less than 3%, is freshwater, and of this, the majority is locked in glaciers, polar ice caps, and deep aquifers. Ensuring fair access to water quantity and quality worldwide is severely hampered by the unequal distribution of water resources. Freshwater availability varies worldwide due to a range of meteorological, geographical, and geological factors. Regions such as the Amazon Basin, the Congo Basin, and Southeast Asia possess abundant renewable water resources.

In contrast, arid and semi-arid areas, including North Africa, the Middle East, and parts of Central Asia, face chronic water scarcity (McDonald et al. 2011). Population growth, urbanisation, and intensive agricultural practices further exacerbate these imbalances, often leading to stress on local water systems. Millions of people remain at risk of waterborne illnesses, particularly in rural and peri-urban regions with poor sanitation infrastructure. The fluctuations in water supply and quality are further complicated by climate change. The hydrological cycle is changing due to melting glaciers, rising global temperatures, shifting precipitation patterns, and more frequent droughts and floods. These changes affect the quantity and quality of water supplies. These changes not only affect ecosystems and biodiversity but also have direct implications for food security, energy production, and human health. A global overview of water quantity and quality distribution thus underscores the urgent need for integrated water resource management, equitable allocation policies, and innovative technological interventions (Van Der Gun 2012).

Water is essential for production, cooling, and processing across industrial sectors, including energy, manufacturing, and mining, making it a major driver of stability and economic growth. In addition, freshwater ecosystems such as rivers, lakes, wetlands, and aquifers support biodiversity, regulate temperature, and provide vital ecosystem services (Hoff 2009). The strategic significance of this limited resource is becoming increasingly evident as water demand rises in tandem with population increase and industrial expansion. In addition to meeting current needs, protecting the world's water resources through equitable distribution, effective management, and technological innovations such as desalination, water recycling, and integrated resource management is essential to guaranteeing resilience and sustainability for future generations (Scanlon et al. 2023).

The objective of this chapter is to provide a comprehensive overview of the global distribution of water resources, covering both quantity and quality. It seeks to examine spatial and temporal variations in freshwater availability across regions of the world, highlighting stark disparities between water-abundant and water-scarce areas. In addition, the chapter aims to analyse the significant issues surrounding water quality by exploring physical, chemical, and biological parameters that influence human health, agricultural productivity, and ecosystem sustainability. It addresses the aspects of water quantity, such as availability, variability, and consumption patterns, as well as water quality, paying particular emphasis to pollution standards, drivers, and their effects on the environment and society. The temporal scope encompasses historical patterns, current circumstances, and future forecasts, influenced by population expansion, socio-economic development, and climatic change (Hoekstra and Hung 2005). The roles of other stakeholders, such as governments, international organizations, businesses, and local communities, whose collective actions shape the outcomes of global water management, are also discussed. By defining these goals and parameters, the chapter aims to provide a comprehensive understanding of the interrelated problems of water scarcity and water quality degradation, as well as insights into solutions that promote sustainability and fair access to this essential resource (Garrote 2017).

1.2 Global Water Resources and Distribution

The literature search initially identified 800 records, which were refined through multi-stage screening based on titles, abstracts, and full-text eligibility. Ultimately, 66 articles met all criteria and were included in the final analysis of global water quality and quantity (Fig. 1.1). Although water covers nearly 71% of Earth's surface, approximately 97.5% is saltwater stored in oceans and seas, unsuitable for direct human consumption, irrigation, or most industrial purposes without desalination. Freshwater comprises only 2.5% of Earth's water, and most of it is not readily accessible. Nearly 68–70% of the world's freshwater is trapped in glaciers, ice caps, and permanent snow, mainly in Greenland and Antarctica. About 30% exists as groundwater, much of which is either non-renewable fossil water accumulated over geological time or too deep for commercial extraction (Baggio et al. 2021). This limited resource must support over eight billion people while meeting agricultural, industrial, and ecosystem needs. Freshwater distribution is highly uneven across regions: arid and semi-arid areas like North Africa, the Middle East, and Central Asia face chronic scarcity, in contrast to water-rich regions such as the Amazon Basin, the Congo, and parts of Canada and Russia. Pollution, rapid population growth, and climate change further strain global freshwater availability and quality.

1.2.1 Global Surface Water Distribution

Surface water, including rivers, lakes, and reservoirs, accounts for less than 1% of Earth's total freshwater yet constitutes the most accessible source for human consumption, agriculture, and industrial use. Rivers contain less than 0.01% of Earth's freshwater, yet they remain essential for water supply, irrigation, hydropower production, and navigation (Zhou et al. 2016). Lakes—including the Great Lakes of North America, Lake Baikal in Russia, and Lake Victoria in Africa—contain a greater percentage of surface water and serve as essential reservoirs for biodiversity, climate regulation, and freshwater storage, supporting millions of people. Reservoirs

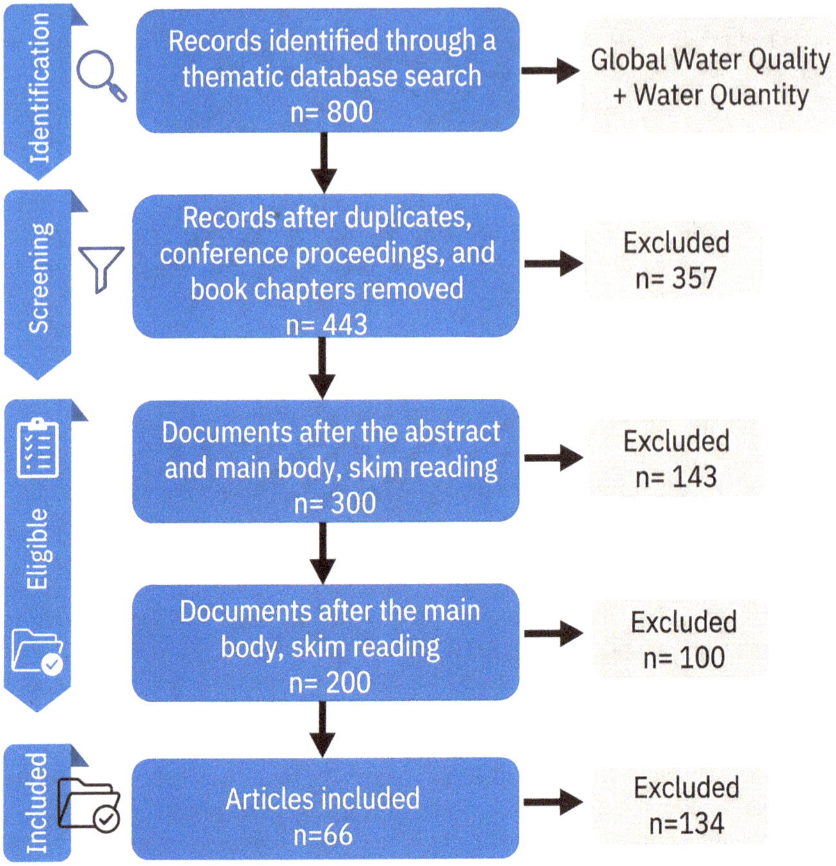

Fig. 1.1 PRISMA flow diagram illustrating the systematic literature search and selection process for global water quality and quantity research. The flowchart details the step-by-step exclusion of records through screening and eligibility phases, resulting in 66 final articles from an initial pool of 800

created by dam construction provide controlled water storage for energy production, agriculture, drinking water supply, and flood control, though they present ecological and socio-economic concerns, including habitat loss, displacement, and altered river flow (Winter 1995). Surface water bodies are disproportionately crucial for maintaining ecosystems and human societies despite their small share of global water resources. Major cities rely on rivers and lakes for transportation, recreation, drinking water, and sanitation. Surface water ecosystems support innumerable species of fish, plants, amphibians, and birds (Biemans et al. 2011). However, untreated sewage discharge and industrial waste deteriorate water quality and endanger human health. Excessive irrigation frequently depletes rivers before they reach the sea, as demonstrated by the Colorado River in North America and the Aral Sea basin in Central Asia (Uddin et al. 2021).

1.2.2 Global Groundwater Distribution

Groundwater is the largest readily available freshwater resource on Earth, accounting for about 30% of the planet's total freshwater resources. It is stored beneath the surface in soil pores, fractures, and geological formations known as aquifers (Hahn and Fuchs 2009). Groundwater provides nearly half of the world's domestic water supply and supports irrigation for more than 40% of global agricultural production. However, excessive extraction, particularly in arid and semi-arid regions, has led to falling water tables, land subsidence, and depletion of non-renewable reserves (Kalbus et al. 2009). Contamination from industrial discharges, agricultural pesticides, and saline intrusion in coastal areas poses growing threats to groundwater quality. Groundwater is essential for climate adaptation, providing resilience during droughts and serving as a natural buffer when rainfall and river flows decrease (McMahon et al. 2011). Yet nitrates from fertilizers, pesticides, heavy metals, and untreated wastewater seeping into aquifers render water unfit for consumption and expensive to treat (Post et al. 2013). Excessive pumping in coastal areas can cause seawater intrusion into freshwater aquifers, salinizing the water and making it unsuitable for drinking and agriculture. Groundwater management has become a global priority due to these challenges (Nouri et al. 2006). Properly managed, groundwater can provide long-term water security for billions; mismanaged, its depletion and contamination could lead to one of the most urgent environmental crises of the future.

1.3 Water Quantity: Availability and Use

The quantity of water available on Earth is vast, but its distribution and usability are highly constrained. Less than 1% of the planet's estimated 1.4 billion cubic kilometers of water are directly accessible for human use in rivers, lakes, and shallow groundwater, and only around 2.5% of it is freshwater. This small fraction must meet the demands of a rapidly growing global population, projected to surpass nine billion by mid-century, while competing with sectors such as agriculture, industry, and domestic consumption. Thus, while the Earth holds more than enough water in absolute terms, the adequate availability of usable freshwater is limited, making sustainable management and equitable distribution critical for meeting present and future needs (Hotloś 2008). To reduce withdrawals and pollution, industries can implement closed-loop systems, cleaner production techniques, and water recycling. To reduce losses and promote household conservation, metropolitan communities must invest in water-sensitive urban design, smart metering, and modern infrastructure. Since many of the world's major rivers, lakes, and aquifers transcend state borders, international collaboration will also be crucial in preventing conflicts through equitable sharing and coordinated management. Table 1.1 and Fig. 1.1 highlight global statistics and studies on water quality, quantity, and distribution.

Table 1.1 Global water quantity and quality distribution

Region	Renewable freshwater resources (m^3/person/year)	% Population under water stress	Total freshwater withdrawals by sector (agriculture/industry/domestic)	Per capita annual water withdrawal (m^3/person)	Access to safely managed drinking water (% population)	Access to safely managed sanitation (% population)	Population lacking safe drinking water (millions)	Waterborne disease burden (annual cases/deaths)	Key water quality issues	Climate vulnerability (water-related risks)	References
World average	~5900	~40%	70% / 20% / 10%	~800	74%	54%	~2000	>1.5 million deaths linked to unsafe water, sanitation, and hygiene	Pathogens, nitrates, industrial effluents, microplastics	Rising floods, droughts, glacier retreat, and aquifer depletion	McDonald et al. (2011)
Africa	~3900	>60%	81% / 6% / 13%	~500	60%	30%	~400	Diarrheal diseases kill ~300,000 children annually	Pathogens, salinity (North Africa), and poor sanitation infrastructure	Severe droughts (Horn of Africa, Sahel), desertification, and climate migration	Jones et al. (2024)
Asia	~3400	~50%	81% / 11% / 8%	~700	82%	64%	~500	Arsenic poisoning in Bangladesh affects >40 million; diarrheal diseases are prevalent in South Asia.	Industrial effluents (India, China), arsenic (Bangladesh), nitrates (Pakistan)	Himalayan glacier melt, extreme monsoon variability, Mekong & Ganges water disputes	Evans et al. (2014); Panday and Kumar (2024)
Europe	~9300	~15%	25% / 55% / 20%	~1200	96%	90%	~20	Low mortality, but emerging contaminants threaten health	Nitrates from agriculture, pharmaceuticals, and heavy metals from industry	Floods in Central & Western Europe, droughts in the Mediterranean and Eastern regions	Williams et al. (2012)
North America	~9600	~25%	40%/45%/15%	~1500	99%	92%	<5	Localised outbreaks; chronic exposure to PFAS, nitrates	PFAS ("forever chemicals"), pharmaceuticals, nutrient runoff, ageing infrastructure	Droughts in the Western U.S., aquifer depletion (Ogallala Aquifer), hurricanes & floods	Scanlon et al. (2007)
South America	~23,000	~20%	65% / 20% / 15%	~1100	86%	78%	~50	Rising diarrheal disease in peri-urban slums; mercury contamination in Amazon mining	Mining pollution (Peru, Chile), untreated wastewater, and deforestation impacts	Amazon deforestation is altering rainfall, glacier retreat in the Andes, and flooding in the lowlands.	Buytaert et al. (2013)
Oceania	~45,000 (Australia: ~23,000)	~30% (higher in Pacific islands)	65% / 15% / 20%	~1700	97%	76%	<5	Isolated outbreaks in Pacific islands	Salinity (Australia), poor infrastructure (Pacific islands), and groundwater contamination	Droughts, sea-level rise, and salinisation of freshwater in small island states	Van Vliet et al. (2021)

The table compares global and regional freshwater resources, water use, access to safe water and sanitation, and related challenges. It shows that water availability and quality vary significantly across regions. Africa and Asia face the most severe water stress and disease burdens due to limited resources, poor infrastructure, and high agricultural dependence. Europe and North America have better water management but face emerging contamination issues, such as nitrates and PFAS. South America and Oceania have abundant resources, yet uneven access and ongoing climate threats, such as droughts, floods, and sea-level rise, persist. (Data corresponds to different years based on the references)

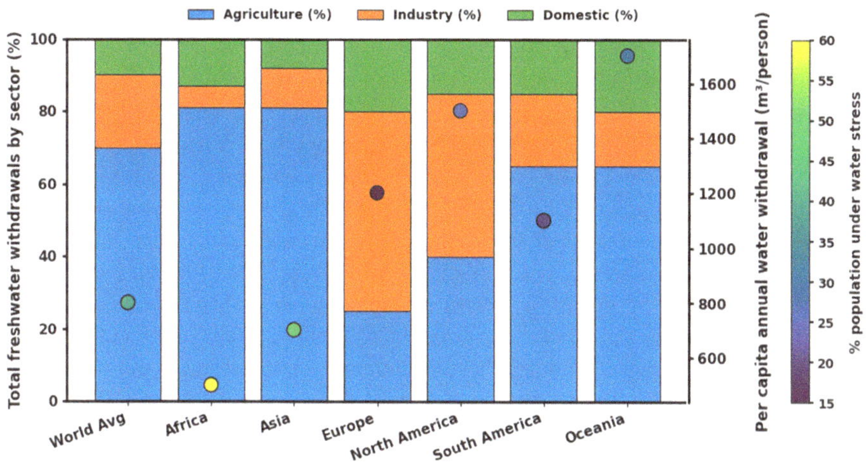

Fig. 1.2 Composite visualization showing regional sectoral freshwater withdrawals (stacked bars), per-capita annual water withdrawal (markers), and the proportion of population under water stress (marker colour)

Figure 1.2 shows how regional water stress emerges from the combined effects of water allocation across sectors and per-person use intensity. Regions dominated by agricultural withdrawals tend to exhibit higher levels of water stress at lower per-capita use, whereas regions with greater industrial shares often sustain higher per-capita withdrawals with comparatively lower stress. This contrast highlights that water stress is not driven solely by withdrawal volumes, but also by sectoral demand patterns and management efficiency.

1.3.1 Dynamics of Global Water Balance

The global water cycle, often called the hydrological cycle, is the continuous flow of water between the atmosphere, land, and seas. Solar energy and gravity are the leading causes. Through evaporation, transpiration, condensation, precipitation, infiltration, and runoff, water is constantly flowing and being redistributed among different reservoirs. Renewable freshwater resources, therefore, depend heavily on climatic conditions and watershed characteristics, as well as human factors such as land use and deforestation, which alter infiltration and runoff patterns (Douville et al. 2021; Panday et al. 2025). Global freshwater demand is dominated by three major sectors: agriculture, industry, and domestic use, each of which places distinct pressures on water resources. Agriculture is by far the largest consumer, accounting for nearly 70% of total freshwater withdrawals worldwide, primarily for irrigation to sustain crop production in regions where rainfall is insufficient. In some arid and semi-arid areas, irrigation is the backbone of food security, yet it often relies on unsustainable groundwater extraction, leading to aquifer depletion and long-term scarcity. Industry represents the second-largest water user, responsible for about 20% of global demand (Chapagain and Hoekstra 2008). Water in this sector is essential for manufacturing, energy generation, cooling in thermal power plants, and processing in industries such as mining, textiles, and chemicals. Industrial water demand is exceptionally high in developed economies, where it often exceeds agricultural use, but it is also rapidly increasing in emerging economies due to industrialization (Flörke et al. 2013).

The balance between availability and demand measures water stress; the Falkenmark Indicator defines stress as <1700 m^3, scarcity as <1000 m^3, and absolute scarcity as <500 m^3 per person/year (Pedro-Monzonís et al. 2015), while the Water Poverty Index (WPI) and Aqueduct Water Risk Atlas provide broader vulnerability data. Climate change redistributes these resources by intensifying the hydrological cycle (Bates et al. 2008), in which initial melt-induced runoff eventually depletes storage and leads to chronic dry-season shortages (Park et al. 2011), exacerbating droughts and increasing reliance on over-extracted groundwater.

1.4 Water Quality: Status and Challenges

Water quality is vital for human health and ecosystems but faces mounting threats from sewage, industrial waste, and agricultural runoff (Evans et al. 2014). Nutrient loading drives eutrophication and harmful blooms, while heavy metals and persistent organics pose bioaccumulation risks. Urbanization fuels microbial contamination, and groundwater is increasingly compromised by arsenic and saltwater intrusion; meanwhile, emerging pollutants like microplastics and pharmaceuticals create new environmental uncertainties (Table 1.2). Climate drivers further disrupt these parameters: warming promotes thermal stratification and algal blooms, intense rainfall mobilizes pathogens and sediment, and droughts increase chemical concentrations, though extreme storms can occasionally dilute nutrient levels (Kumar et al. 2025). These climate-driven hydrological shifts fundamentally reshape water safety and ecosystem health worldwide (Silori et al. 2024).

Water quality refers to the physical, chemical, and biological properties that determine a water body's suitability for drinking, irrigation, and ecosystem support. Physical parameters, such as temperature, color, and turbidity, influence both aesthetic acceptance and ecological functions, including light penetration and dissolved oxygen levels (Akhtar et al. 2021). Chemical indicators—including pH, salinity, and nutrient levels such as nitrogen—describe the balance of dissolved substances. While these occur naturally, human-induced excesses lead to eutrophication and the introduction of toxic heavy metals and pesticides, which jeopardize health and biodiversity (Boyd 2000). Anthropogenic activity is the primary driver of global water degradation. Industrial sectors introduce heavy metals and thermal waste that bioaccumulate, while agriculture contributes fertilizers and pesticides that cause nutrient-driven algal blooms and oxygen depletion (Popescu and Ionel 2010). Domestically, untreated sewage and greywater introduce pathogens into

Table 1.2 Water quality variations driven by climate across global river and lake systems

Location	Study period	Key climate driver	Observation on water quality	Impact category	References
Danube River, Bratislava, Slovakia	1926–2005	Rising air temperature	Long-term warming influenced river temperatures, but annual discharge variability moderated the overall warming trend.	Temperature-driven	Pekarova et al. (2008)
Lakes Purrumbete, Colac & Bullen Merri, Australia	1984–2000	Temperature + rainfall deficit	Increasing salinity and nutrient enrichment are linked to a warmer climate and reduced rainfall.	Multi-driver	Tibby and Tiller (2007)
Lake Tahoe, USA	1970–2007	Temperature rise	Higher thermal stratification reduced oxygen availability in deeper layers.	Temperature-driven	Sahoo et al. (2010)
Neuse River Estuary, USA	1979–2003	Extreme storms/ hurricanes	Storm-induced flushing temporarily lowered eutrophication and reduced algal bloom risk.	Extreme-event-driven	Paerl et al. (2006); Paerl and Huisman (2008)
River Meuse, Western Europe	1976–2003	Drought + warming	Low flows and higher temperatures increased dissolved concentrations of elements and metals, leading to nutrient-driven algal blooms.	Multi-driver	Van Vliet and Zwolsman (2008)

(continued)

Table 1.2 (continued)

Location	Study period	Key climate driver	Observation on water quality	Impact category	References
Lake Taihu, China	2007	Unusually warm spring	Warm conditions triggered an extreme cyanobacterial bloom, causing Microcystis toxin contamination and disrupting the drinking water supply.	Temperature-driven	Qin et al. (2010)
Sau Reservoir, Spain	1964–2007	Streamflow changes	Variations in inflow had a more substantial effect on dissolved oxygen depletion than did temperature.	Flow-driven	Marcé et al. (2010)
22 upland waters, UK	1988–2002	Temperature + rainfall + land-use	DOM concentrations increased due to warming, wetness, and other catchment factors, including CO_2 and acid deposition.	Multi-driver	Evans et al. (2005)
Coastal rivers, Western Finland	1913–2007	High rainfall	Acidic pulses (low pH) are observed during wet periods in acid sulfate soil catchments.	Precipitation-driven	Saarinen et al. (2010)

Table 1.2 (continued)

Location	Study period	Key climate driver	Observation on water quality	Impact category	References
15 mountain rivers, Northern Spain	1973–2005	Temperature rise	Warmer conditions are linked to elevated nutrient and dissolved organic carbon levels in semi-arid systems.	Temperature-driven	Benítez-Gilabert et al. (2010)
30 coastal rivers & groundwater, Western France	1973–2007	Rainfall + temperature	Nutrient levels fluctuated with climatic variability; groundwater showed a delayed response.	Multi-driver	Gascuel-Odoux et al. (2010)
Girnock Catchment, Scotland	14 months	High rainfall	Fecal contamination risks increased during rainy months.	Precipitation-driven	Tetzlaff et al. (2010)
27 rivers, Japan	1987–1995	Warmer & wetter conditions	Organic matter, sediments, and COD increased with rainfall and temperature; DO declined with warming.	Multi-driver	Ozaki et al. (2003)
Conestoga River Basin, USA	1977–1997	Annual rainfall	Total nitrogen loads strongly corresponded to rainfall variability.	Precipitation-driven	Chang (2004)
USA (multi-site)	1948–1994	Intense rainfall	Higher runoff during heavy rains is linked with outbreaks of waterborne diseases.	Extreme-event-driven	Taylor et al. (2009)

(continued)

Table 1.2 (continued)

Location	Study period	Key climate driver	Observation on water quality	Impact category	References
Northern & Eastern Uganda	1999–2001; 2004; 2007	Rainy season	Fecal coliform levels rose in groundwater-fed sources during wet periods.	Precipitation-driven	Tumwine et al. (2002)
Taiwan, China	1998	Heavy rainfall	The probability of enterovirus infection increased substantially during high-intensity rainfall (>31 mm h^{-1}).	Extreme-event-driven	Jean et al. (2006)
Rhine Basin	1980–2001	Seasonal precipitation	Nutrient concentrations varied seasonally with rainfall-driven soil erosion.	Precipitation-driven	Loos et al. (2009)
River Thames, England	1868–2008	Changes in runoff + land use	Nutrient loads linked with variability in runoff and anthropogenic land-use modifications.	Multi-driver	Howden et al. (2010)

This table presents a comprehensive synthesis of how water quality in global freshwater bodies responds to key climate-related drivers, including rising air and water temperatures, altered precipitation regimes, drought intensity, storm events, and changes in river discharge

rivers, causing widespread waterborne illnesses like cholera, particularly in developing nations. Urban stormwater further adds plastics and oils to these systems, requiring integrated pollution control and improved wastewater infrastructure to safeguard resources for future generations.

Figure 1.3 synthesizes global water-quality distribution as a function of water quantity constraints rather than solely pollutant presence. Regions with limited freshwater availability and reduced flushing capacity are dominated by concentration-driven quality degradation, whereas agriculturally intensive regions exhibit nutrient and organic enrichment when assimilative capacity is seasonally exceeded. Highly industrialised regions show chemical accumulation associated with high per-capita withdrawals and regulated flows, while pathogen-dominated regimes prevail where water quantity is insufficient to sustain safe sanitation and hygiene. The figure highlights that global water-quality patterns are governed by the interaction between

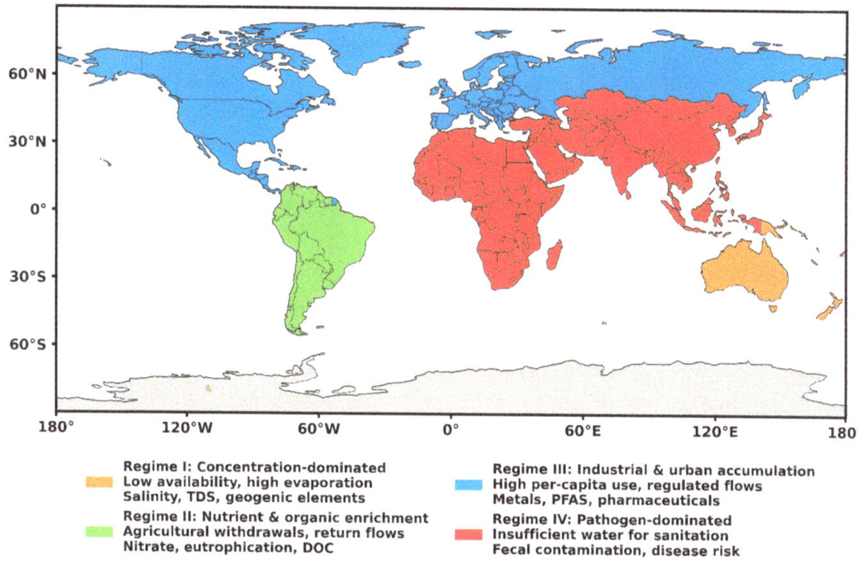

Fig. 1.3 Global map illustrating dominant water-quality regimes emerging from the spatial imbalance between freshwater availability, human water use, and assimilative capacity. Colours indicate prevailing regime types, while the legend summarises the controlling hydrological context and characteristic water-quality issues

availability, use, and natural assimilation capacity, explaining why similar pollutant loads produce different outcomes across regions.

Modern resources are increasingly threatened by emerging contaminants like microplastics and pharmaceuticals, which pose poorly understood risks. Microplastics from synthetic textiles bioaccumulate in food chains, while trace pharmaceuticals—including antibiotics and hormones—fuel concerns regarding antibiotic resistance and endocrine disruption (Kinigopoulou et al. 2022). Persistent heavy metals, such as arsenic and mercury from mining or improper waste disposal, remain critical because they do not degrade and continue to poison groundwater for generations. Managing these synergistic threats requires advanced treatment technologies and global monitoring frameworks (Agarwal et al. 2024).

1.4.1 Regional Perspectives in Water Quantity and Quality

In many rural and peri-urban areas of Africa, limited access to safe water forces reliance on unprotected sources, leading to frequent outbreaks of cholera and dysentery (Falkenmark 1989). Water quality is further degraded by heavy metals from artisanal mining, industrial pollution, and agricultural runoff. Poor sanitation results in untreated sewage discharge into water bodies, while vital groundwater reserves face threats from salinization, fluoride contamination, and over-extraction

(Livingston 2021). Asia faces severe water degradation driven by rapid industrialization and agricultural intensification. Excessive fertilizer use causes widespread eutrophication, while major rivers such as the Ganges and Citarum are heavily contaminated by industrial effluents and thermal pollution (Cropper and Griffiths 1994). Untreated sewage fuels microbial contamination in dense urban centers, and groundwater in South and Southeast Asia is increasingly compromised by arsenic and fluoride. Balancing economic growth with water conservation requires stricter regulation and sustainable farming practices.

Despite robust governance, Europe faces regional water stress; Southern Europe struggles with low rainfall and high agricultural demand, while Northern regions face industrial and urban effluents (Kallioras et al. 2006). Emerging threats like microplastics and pharmaceutical residues are testing current frameworks. However, Europe's management of transboundary rivers—such as through the Danube (ICPDR) and Rhine (ICPR) commissions—serves as a global model for regional cooperation and institutional governance. North America possesses vast freshwater reserves, including the Great Lakes and the Ogallala Aquifer, providing high per capita availability (Hillstrom and Hillstrom 2003). However, this abundance is offset by uneven distribution and overuse. In the Western U.S. and Mexico, intensive agricultural and urban demands have led to the depletion of the Ogallala Aquifer and the severe diversion of the Colorado River, which now rarely reaches its delta (McKinley and Johnston 2010).

1.4.2 Technological and Innovative Solutions

Digital technologies such as remote sensing, GIS, and satellite imagery are transforming water management by enhancing forecasting and early warning systems. In urban settings, smart meters and sensors improve supply efficiency by identifying leaks and enabling demand-based pricing (Ahmed et al. 2023). Nature-based solutions, including constructed wetlands and green infrastructure, offer climate resilience, while advancements in nanotechnology and biotechnology provide new pathways to remove heavy metals and pharmaceuticals. Despite their potential, widespread adoption is often hindered by high costs and gaps in technical capacity. Water treatment has been revolutionized by advanced membrane technologies—such as nanofiltration and reverse osmosis—which are increasingly replacing conventional sand filtration (Adesina et al. 2024). These processes, along with Advanced Oxidation Processes (AOPs) and UV irradiation, effectively remove pathogens and degrade organic contaminants like endocrine disruptors.

Desalination, primarily utilizing reverse osmosis (RO), now provides essential water security in arid regions due to declining costs and improved energy efficiency (Matin et al. 2021). However, managing highly saline brine and the substantial energy requirements remain significant environmental challenges (Wencki et al. 2020). Furthermore, IoT-enabled smart infrastructure integrates real-time sensors to monitor flow, pressure, and contamination across distribution networks. This reduces non-revenue water losses and ensures safer supplies, while smart meters encourage household conservation through real-time usage feedback (Overmars and Venkatraman 2020). When paired with AI, these systems facilitate predictive maintenance and long-term demand forecasting, though cybersecurity threats and high installation costs hinder implementation (Kamienski et al. 2019).

1.5 Conclusion

Earth's water paradox lies in the gap between total volume and accessible supply, which is unevenly distributed and further constrained by seasonal variability, rising demand, and climate change. Simultaneously, quality is declining due to industrial effluents, agricultural runoff, and emerging contaminants like microplastics, threatening both human health and ecosystems. Regional disparities underscore that scarcity is a product of both natural endowment and socio-economic factors; while Africa and South Asia face chronic shortages, North America and Europe manage distinct contamination challenges. Addressing these interconnected challenges requires integrated, forward-looking strategies. Technological innovations in treatment and monitoring are promising but must be paired with robust governance, equitable pricing, and community-led management. Holistic approaches—including nature-based solutions and transboundary cooperation—are essential to balance human needs with ecological health. As water is central to the global sustainability agenda, achieving long-term security requires a commitment to science-based stewardship and international cooperation to safeguard this vital resource for future generations.

References

Adesina OB, William C, Oke EI (2024) Evolution in water treatment: exploring traditional self-purification methods and emerging technologies for drinking water and wastewater treatment: a review. World News Nat Sci 53:169–185

Agarwal V, Kumar M, Panday DP, Zang J, Munoz-Arriola F (2024) Unlocking the potential of remote sensing for arsenic contamination detection and management: challenges and perspectives. Curr Opin Environ Sci Health 100578. https://doi.org/10.1016/j.coesh.2024.100578

Ahmed F, Johnson D, Hashaikeh R, Hilal N (2023) Barriers to innovation in water treatment. Water 15(4):773

Akhtar N, Syakir Ishak MI, Bhawani SA, Umar K (2021) Various natural and anthropogenic factors responsible for water quality degradation: a review. Water 13(19):2660

Baggio G, Qadir M, Smakhtin V (2021) Freshwater availability status across countries for human and ecosystem needs. Science of the Total Environment 792:148230

Bates B, Kundzewicz Z, Wu S (2008) Climate change and water. Intergovernmental Panel on Climate Change Secretariat

Benítez-Gilabert M, Alvarez-Cobelas M, Angeler DG (2010) Effects of climatic change on stream water quality in Spain. Clim Chang 103(3):339–352

Biemans H, Haddeland I, Kabat P, Ludwig F, Hutjes RW, Heinke J et al (2011) Impact of reservoirs on river discharge and irrigation water supply during the 20th century. Water Resour Res 47(3):W03509

Boyd CE (2000) Water quality: an introduction. Springer Science & Business Media

Buytaert W, Breuer L (2013) Water resources in South America: sources and supply, pollutants and perspectives. Proceedings of the IAHS-IAPSO-IASPEI Assembly, Gothenburg, Sweden 106–113

Chang H (2004) Water quality impacts of climate and land use changes in southeastern Pennsylvania. Prof Geogr 56(2):240–257

Chapagain AK, Hoekstra AY (2008) The global component of freshwater demand and supply: an assessment of virtual water flows between nations as a result of trade in agricultural and industrial products. Water Int 33(1):19–32

Cropper M, Griffiths C (1994) The interaction of population growth and environmental quality. Am Econ Rev 84(2):250–254

Douville H, Raghavan K, Renwick J, Allan RP, Arias PA, Barlow M et al (2021) Water cycle changes. IPCC

Evans CD, Monteith DT, Cooper DM (2005) Long-term increases in surface water dissolved organic carbon: observations, possible causes and environmental impacts. Environ Pollut 137(1):55–71

Evans AE, Hanjra MA, Jiang Y, Qadir M, Drechsel P (2014) Water quality: assessment of the current situation in Asia. In: Water quality policy and management in Asia. Routledge, pp 9–30

Falkenmark M (1989) The massive water scarcity now threatening Africa: why isn't it being addressed? Ambio 18:112–118

Flörke M, Kynast E, Bärlund I, Eisner S, Wimmer F, Alcamo J (2013) Domestic and industrial water uses of the past 60 years as a mirror of socio-economic development: a global simulation study. Glob Environ Chang 23(1):144–156

Garrote L (2017) Managing water resources to adapt to climate change: facing uncertainty and scarcity in a changing context. Water Resour Manag 31(10):2951–2963

Gascuel-Odoux C, Aurousseau P, Durand P, Ruiz L, Molenat J (2010) The role of climate on inter-annual variation in stream nitrate fluxes and concentrations. Sci Total Environ 408(23):5657–5666

Hahn HJ, Fuchs A (2009) Distribution patterns of groundwater communities across aquifer types in South-Western Germany. Freshw Biol 54(4):848–860

Hillstrom K, Hillstrom LC (2003) North America: a continental overview of environmental issues. Bloomsbury Publishing USA

Hoekstra AY, Hung PQ (2005) Globalisation of water resources: international virtual water flows in relation to crop trade. Glob Environ Chang 15(1):45–56

Hoff H (2009) Global water resources and their management. Curr Opin Environ Sustain 1(2):141–147

Hotloś HALINA (2008) Quantity and availability of freshwater resources: the world, Europe, and Poland. Environ Prot Eng 34(2):67–77

Howden NJK, Burt TP, Worrall F, Whelan MJ, Bieroza M (2010) Nitrate concentrations and fluxes in the River Thames over 140 years (1868–2008): are increases irreversible? Hydrol Process 24(18):2657–2662

Jean J-S, Guo H-R, Chen S-H, Liu C-C, Chang W-T, Yang Y-J, Huang M-C (2006) The association between rainfall rate and occurrence of an enterovirus epidemic due to a contaminated well. J Appl Microbiol 101(6):1224–1231

Jones ER, Bierkens MF, van Vliet MT (2024) Current and future global water scarcity intensifies when accounting for surface water quality. Nature Climate Change 14(6):629–635

Kalbus E, Schmidt C, Molson JW, Reinstorf F, Schirmer M (2009) Influence of aquifer and streambed heterogeneity on the distribution of groundwater discharge. Hydrol Earth Syst Sci 13(1):69–77

Kallioras A, Pliakas F, Diamantis I (2006) The legislative framework and policy for water resources management of transboundary rivers in Europe: the case of the Nestos/Mesta River between Greece and Bulgaria. Environ Sci Pol 9(3):291–301

Kamienski C, Soininen JP, Taumberger M, Dantas R, Toscano A, Salmon Cinotti T et al (2019) Smart water management platform: IoT-based precision irrigation for agriculture. Sensors 19(2):276

Kinigopoulou V, Pashalidis I, Kalderis D, Anastopoulos I (2022) Microplastics as carriers of inorganic and organic contaminants in the environment: a review of recent progress. J Mol Liq 350:118580

Kumar M, Panday DP, Goswami R, Neog N, Bahukhandi KD, Zhang J, Bhattacharya P (2025) Dynamics of arsenic co-contamination in the groundwater of the Eastern Himalayas: the biogeochemical drivers and GRACE-based evidence. ACS ES & T Water. https://doi.org/10.1021/acsestwater.4c00106

Livingston J (2021) Water scarcity & health in urban Africa. Daedalus 150(4):85–102

Loos S, Middelkoop H, van der Perk M, van Beek R (2009) Large-scale nutrient modelling using globally available datasets: a test for the Rhine basin. J Hydrol 369(3–4):403–415

Marcé R, Rodríguez-Arias MT, García JC, Armengol J (2010) El Niño Southern Oscillation and climate trends impact reservoir water quality. Glob Chang Biol 16(10):2857–2865

Matin A, Laoui T, Falath W, Farooque M (2021) Fouling control in reverse osmosis for water desalination & reuse: current practices & emerging environment-friendly technologies. Sci Total Environ 765:142721

McDonald RI, Douglas I, Revenga C, Hale R, Grimm N, Grönwall J, Fekete B (2011) Global urban growth and the geography of water availability, quality, and delivery. Ambio 40(5):437–446

McKinley A, Johnston EL (2010) Impacts of contaminant sources on marine fish abundance and species richness: a review and meta-analysis of evidence from the field. Mar Ecol Prog Ser 420:175–191

McMahon PB, Plummer LN, Böhlke JK, Shapiro SD, Hinkle SR (2011) A comparison of recharge rates in aquifers of the United States based on groundwater-age data. Hydrogeol J 19(4):779–800

Nouri J, Mahvi AH, Babaei A, Ahmadpour E (2006) Regional pattern distribution of groundwater fluoride in the Shush aquifer of Khuzestan County, Iran. Fluoride 39(4):321

Overmars A, Venkatraman S (2020) Towards a secure and scalable IoT infrastructure: a pilot deployment for a smart water monitoring system. Technologies 8(4):50

Ozaki N, Fukushima T, Harasawa H, Kojiri T, Kawashima K, Ono M (2003) Statistical analyses on the effects of air temperature fluctuations on river water qualities. Hydrol Process 17(14):2837–2853

Paerl HW, Huisman J (2008) Climate—blooms like it hot. Science 320(5872):57–58

Paerl HW, Valdes LM, Piehler MF, Stow CA (2006) Assessing the effects of nutrient management in an estuary experiencing climatic change: the Neuse River Estuary, North Carolina. Environ Manag 37(3):422–436

Panday DP, Kumar M (2024) Decoding uranium variation over the Indian Peninsula through leveraging standardized precipitation evapotranspiration indices and groundwater level fluctuations. Groundw Sustain Dev:101241. https://doi.org/10.1016/J.GSD.2024.101241

Panday DP, Kumari A, Kumar M (2025) Alkalinity-salinity-sustainability: decadal groundwater trends and its impact on agricultural water quality in the Indian Peninsula. Sci Total Environ 978:179459. https://doi.org/10.1016/j.scitotenv.2025.179459

Park JY, Park MJ, Ahn SR, Park GA, Yi JE, Kim GS et al (2011) Assessment of future climate change impacts on water quantity and quality for a mountainous dam watershed using SWAT. Trans ASABE 54(5):1725–1737

Pedro-Monzonís M, Solera A, Ferrer J, Estrela T, Paredes-Arquiola J (2015) A review of water scarcity and drought indices in water resources planning and management. J Hydrol 527:482–493

Pekarova P, Halmova D, Miklanek P, Onderka M, Pekar J, Skoda P (2008) Is the water temperature of the Danube River at Bratislava, Slovakia, rising? J Hydrometeorol 9(5):1115–1122

Popescu F, Ionel I (2010) Anthropogenic air pollution sources. In: Air quality. InTech Open, pp 1–22
Post VE, Vandenbohede A, Werner AD, Teubner MD (2013) Groundwater ages in coastal aquifers. Adv Water Resour 57:1–11
Qin B, Zhu G, Gao G, Zhang Y, Li W, Paerl HW, Carmichael WW (2010) A drinking water crisis in Lake Taihu, China: linkage to climatic variability and lake management. Environ Manag 45(1):105–112
Saarinen T, Vuori K-M, Alasaarela E, Kløve B (2010) Long-term trends and variation of acidity, CODMn and colour in coastal rivers of western Finland in relation to climate and hydrology. Sci Total Environ 408(21):5019–5027
Sahoo GB, Schladow SG, Reuter JE, Coats R (2010) Effects of climate change on thermal properties of lakes and reservoirs, and possible implications. Stoch Env Res Risk A 25(4):445–456
Scanlon BR, Fakhreddine S, Rateb A, de Graaf I, Famiglietti J, Gleeson T et al (2023) Global water resources and the role of groundwater in a resilient water future. Nat Rev Earth Environ 4(2):87–101
Scanlon BR, Jolly I, Sophocleous M, Zhang L (2007) Global impacts of conversions from natural to agricultural ecosystems on water resources: Quantity versus quality. Water resources research 43(3)
Silori R, Agarwal V, Panday DP, Giri BS, Khandelwal M, Garg M, Shrivastava V (2024) Assessing groundwater quality in the rapidly Industrialising Selaqui Area of Dehradun, India: implications for sustainable water management. Water Conserv Sci Eng 9(2):1–12. https://doi.org/10.1007/s41101-024-00320-5
Taylor R, Miret-Gaspa M, Tumwine J, Mileham L, Flynn R, Howard G, Kulabako R (2009) Increased risk of diarrhoeal diseases from climate change: evidence from communities supplied by groundwater in Uganda. In: Taylor R, Tindimugaya C, Owor M, Shamsudduha M (eds) Groundwater and climate in Africa. Proceedings of the Kampala Conference, 24–28 June 2008, International Association of Hydrological Sciences (IAHS). IAHS Publication No. 334, IAHS Press, Wallingford, pp 15–19
Tetzlaff D, Soulsby C, Birkel C (2010) Hydrological connectivity and microbiological fluxes in montane catchments: the role of seasonality and climatic variability. Hydrol Process 24(9):1231–1235
Tibby J, Tiller D (2007) Climate-water quality relationships in three western Victorian (Australia) lakes 1984–2000. Hydrobiologia 591(1):219–234
Tumwine J, Kekitiinwa A, Nabukeera N, Akiyoshi D, Buckholt M, Tzipori S (2002) Enterocytozoon bieneusi among children with diarrhea attending Mulago Hospital in Uganda. Am J Trop Med Hyg 67(3):299–303
Uddin MG, Nash S, Olbert AI (2021) A review of water quality index models and their use for assessing surface water quality. Ecol Indic 122:107218
Van Der Gun J (2012) The United Nations world water development report: groundwater and global change: trends, opportunities and challenges. UNESCO
Van Vliet MT, Jones ER, Flörke M, Franssen WH, Hanasaki N, Wada Y, Yearsley JR (2021) Global water scarcity including surface water quality and expansions of clean water technologies. Environmental Research Letters 16(2):024020
Van Vliet MTH, Zwolsman JJG (2008) Impact of summer droughts on the water quality of the Meuse river. J Hydrol 353(1–2):1–17
Wencki K, Thöne V, Ante A, Hogen T, Hohmann C, Tettenborn F et al (2020) Approaches for the evaluation of future-oriented technologies and concepts in the field of water reuse and desalination. J Water Reuse Desalin 10(4):269–283
Williams R, Keller V, Voß A, Bärlund I, Malve O, Riihimäki J, Alcamo J (2012) Assessment of current water pollution loads in Europe: estimation of gridded loads for use in global water quality models. Hydrological processes 26(16):2395–2410
Winter TC (1995) Recent advances in understanding the interaction of groundwater and surface water. Rev Geophys 33(S2):985–994
Zhou T, Nijssen B, Gao H, Lettenmaier DP (2016) The contribution of reservoirs to global land surface water storage variations. J Hydrometeorol 17(1):309–325

Chapter 2
Transboundary Water Management and Sharing

Abstract Transboundary freshwater systems are critical ecological and political assets that sustain billions of people while also serving as focal points of persistent international disputes. Governing these shared waters poses a multidimensional challenge at the intersection of environmental management, law, and diplomacy, and many basins remain vulnerable to political tension, power asymmetry, and environmental stress despite decades of cooperative frameworks. Understanding the structural and strategic determinants of cooperation has therefore become central to hydropolitical research and policy. This study reviews global transboundary water interactions using data from the Transboundary Freshwater Political Process Database (TFPD) and the Basins at Risk (BAR) Scale, which quantifies cooperative and conflictive events among riparian states from 1948 to 2008. Empirical patterns are integrated with game-theoretic frameworks that conceptualize hydropolitical negotiations as strategic interactions shaped by incentives, trust, and enforcement capacity. Spatial analysis reveals pronounced geopolitical clustering, with high conflict intensity (−5 to −7) concentrated in the Middle East, South Asia, and the Horn of Africa, and strong cooperation (+5 to +7) characterizing basins in Europe and North America. Temporally, cooperative behavior increases markedly after 1990, coinciding with the expansion of multilateral basin organizations and legal mechanisms. The synthesis demonstrates that cooperation arises not from hydrological abundance, but from institutional depth, transparency, and adaptive governance. Game-theoretic analysis further shows that cooperative Nash equilibria emerge when institutions internalize transaction costs, foster trust, and enable equitable benefit-sharing, offering predictive insight into basins at risk of conflict. Overall, the convergence of quantitative event data and strategic negotiation modeling provides a robust foundation for preventive diplomacy and resilient transboundary water governance, enabling shared rivers to evolve from arenas of competition into platforms for long-term cooperation and peacebuilding.

Keywords Transboundary water · Hydro-diplomacy · Cooperation · Institutional governance · Water treaties

M. Kumar, D. P. Panday, *Enabling Water Security and Sustainability: Global Challenges and Solutions*, SpringerBriefs in Water Science and Technology, https://doi.org/10.1007/978-3-032-21521-5_2

2.1 Introduction

Water is an essential resource for sustaining life, ecosystems, and socio-economic development. Globally, freshwater resources are unevenly distributed, both spatially and temporally, creating challenges for access and utilization. A significant portion of the world's rivers, lakes, and aquifers cross national boundaries, forming **transboundary water resources**. These shared waters are critical not only for ecological and economic sustainability but also for maintaining peace and regional stability. Transboundary water management refers to the coordinated planning, allocation, and governance of water resources shared between two or more sovereign states, ensuring equitable use while minimizing conflict and environmental degradation (Wolf et al. 2003; Dinar et al. 2007).

Throughout history, water has been both a source of cooperation and contention. Ancient civilizations, such as those in Mesopotamia and the Nile Valley, recognized the importance of regulating shared waters to support agriculture and settlements (Allan 2001). However, even in modern times, disputes over water allocation persist due to upstream-downstream asymmetries, differences in economic development, and political tensions. For example, the Nile River, shared by 11 countries, has historically been dominated by Egypt due to its reliance on Nile waters, leading to tensions with upstream states such as Ethiopia, especially over recent projects such as the Grand Ethiopian Renaissance Dam (Swain 2011). Similarly, the Indus River system has been a source of both conflict and collaboration between India and Pakistan, with the 1960 Indus Waters Treaty emerging as a landmark agreement in conflict resolution (Kerr 2012). Conflicts over shared waters typically arise from three key factors: water scarcity, inequitable allocation, and unilateral development. Upstream countries often seek to maximize their benefits from hydropower, irrigation, or urban water supply, while downstream states are concerned about reduced flows, deteriorating water quality, and ecological impacts. These disputes are exacerbated by climate variability, population growth, and industrial demands, highlighting the urgent need for coordinated management strategies (Gleick 1993).

The increasing frequency of water-related conflicts has underscored the importance of proactive management of shared water resources. Effective transboundary water management ensures equitable, sustainable, and environmentally sensitive water allocation. It provides mechanisms to prevent disputes, mitigate risks of water scarcity, and promote regional cooperation. Beyond resource allocation, management frameworks also address ecological sustainability, disaster risk reduction (floods and droughts), and socio-economic development (UN Water 2020). Failure to adopt cooperative management strategies can escalate into international tensions, economic losses, and environmental degradation.

Recognizing the potential for conflict and the necessity of cooperation, the international community has developed legal principles and treaties to govern shared water resources. The 1997 United Nations Convention on the Law of the Non-Navigational Uses of International Watercourses (UN Watercourses Convention) establishes norms for equitable utilization, prevention of significant harm, and prior notification of planned measures affecting other states (UN 1997). Earlier frameworks, such as the Helsinki Rules (1966) and the Berlin Rules (2004), set principles for sustainable and equitable management of transboundary waters, influencing national and regional policies (McCaffrey 2007). Region-specific agreements, including the Southern African Development Community (SADC) Protocol on Shared Watercourses and the European Union Water Framework Directive, operationalize these principles, facilitating cooperation and joint management among riparian states (Earle and Bazilli 2008).

Despite these frameworks, challenges persist due to differences in national priorities, institutional capacities, and enforcement mechanisms. To complement legal strategies, analytical tools such as game theory have been increasingly applied to model negotiation dynamics, predict outcomes, and design incentive-compatible strategies for conflict resolution. Game-theoretic approaches enable policymakers to identify mutually beneficial solutions while accounting for strategic behavior, enhancing the effectiveness of transboundary water governance (Dinar 2003). This chapter aims to provide a comprehensive understanding of transboundary water management and sharing in a global context. The specific objectives of this study are to provide a historical overview of transboundary water management trends through case studies that illustrate the evolution of disputes and cooperative frameworks, while examining the global laws and principles—including international legal instruments and regional agreements—that govern these shared resources. Furthermore, the research assesses the efficacy of game-theoretic approaches in conflict resolution, evaluating how strategic modeling contributes to the development of effective and equitable water-sharing mechanisms.

2.2 Evolution of Trends in Transboundary Water Management

Transboundary water resources have historically been both a source of cooperation and a source of conflict, reflecting the tension between shared dependence and competing national interests. These resources, encompassing rivers, lakes, and aquifers that traverse political boundaries, have required evolving management strategies to balance ecological sustainability, economic development, and geopolitical stability. Historical and contemporary case studies provide insights into how different regions have addressed these challenges and how cooperative frameworks have emerged over time.

The concept of shared water management is ancient. Evidence from Mesopotamia, the Nile Valley, and the Indus civilization demonstrates early regulatory efforts to prevent disputes over water use. In Mesopotamia, irrigation agreements dating back to the nineteenth century BCE codified equitable water distribution among agricultural communities (Allan 2001). In Egypt, the Nile's predictable inundation cycles were central to agrarian productivity, necessitating centralized control and cooperative management across settlements. Similarly, in the Indus Valley, early hydraulic infrastructure, including canals and reservoirs, required coordination among neighbouring city-states to ensure equitable access (Lowi 1993). These historical examples highlight two critical themes: water scarcity often catalyzes conflict, and even in antiquity, societies sought institutional mechanisms for cooperation.

The twentieth century introduced new dynamics to transboundary water management. Nation-state formation, colonial legacies, and the intensification of agricultural, industrial, and urban demands created upstream-downstream asymmetries, often resulting in conflicts. Upstream countries frequently pursued hydropower and irrigation projects that altered flow regimes, while downstream states experienced reduced water availability and ecological degradation (Turton 2003). The Nile, spanning 11 countries, has historically been dominated by Egypt, which relies almost entirely on its waters for agriculture and domestic supply. The 1929 Nile Waters Agreement and the 1959 Egypt-Sudan Treaty formalized water allocation in favor of downstream states, excluding upstream nations such as Ethiopia (Swain 2011). Ethiopia's construction of the Grand Ethiopian Renaissance Dam (GERD) has reignited tensions, emphasizing the enduring upstream-downstream asymmetry and the importance of cooperative frameworks that account for all riparian states. The Indus River, shared by India and Pakistan, exemplifies the potential for both conflict and cooperation. Historical disputes over allocation were mitigated through the Indus Waters Treaty of 1960, mediated by the World Bank. The treaty allocated water resources between the two countries and established mechanisms for joint management and dispute resolution, thereby illustrating the effectiveness of third-party facilitation in transboundary water governance (Kerr 2012).

The Mekong River traverses China, Myanmar, Laos, Thailand, Cambodia, and Vietnam. Rapid development in upstream regions, including dam construction in China and Laos, has disrupted ecosystems and caused downstream water stress. The Mekong River Commission (MRC), established in 1995, represents a cooperative governance model; however, its effectiveness is constrained by limited enforcement powers and differing national priorities (Dore 2003). This case underscores the necessity of robust institutional mechanisms and regional coordination in complex multi-state river systems. Shared by Israel, Jordan, Palestine, Lebanon, and Syria, the Jordan River has historically been a flashpoint due to its scarcity and strategic significance. Early unilateral development efforts by riparian states generated conflicts. Still, the **1994** Israel-Jordan Peace Treaty included detailed water-sharing arrangements, highlighting how transboundary water governance can be integrated into broader peace and security frameworks (Wolf 1995).

The timeline (Fig. 2.1) demonstrates the historical evolution of transboundary water management from early efforts focused on navigation and boundary regulation to increasingly complex institutional, legal, and cooperative frameworks. Early twentieth-century treaties emphasized allocation and dispute resolution, while the post-war period marked the emergence of basin organizations and principles of equitable and reasonable use. From the 1990s onward, cooperation expanded through multilateral conventions, basin-wide initiatives, and systematic assessment

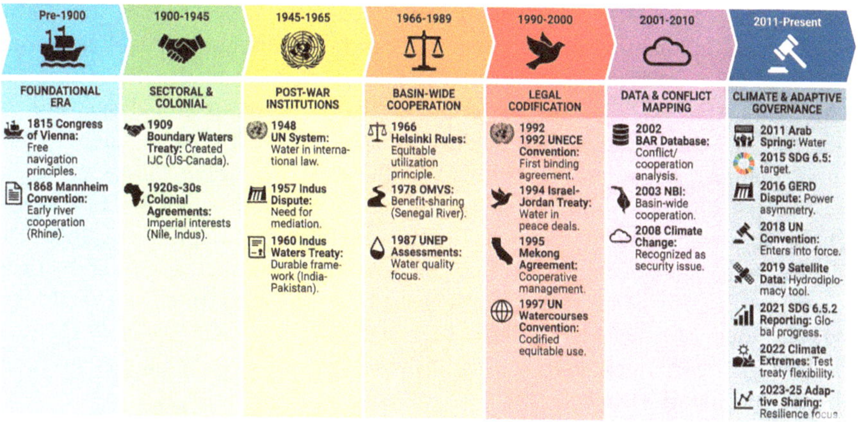

Fig. 2.1 Timeline of major global milestones in transboundary water management and sharing (1815–2025), highlighting the evolution from navigation-oriented agreements to basin-wide cooperation, legal codification, data-driven analysis, and climate-adaptive governance

of conflict and cooperation, culminating in contemporary approaches that integrate sustainability goals, data transparency, and climate adaptation. Together, these milestones reflect a gradual shift from static, treaty-based arrangements to dynamic and adaptive governance systems designed to manage uncertainty, asymmetry, and shared risk in transboundary river basins.

Analysis of these historical and contemporary cases reveals several significant trends in the evolution of transboundary water governance. First, there has been a marked shift from conflict-driven interactions toward cooperative mechanisms, reflecting the recognition that shared water resources require joint management to ensure sustainability. Second, the institutionalization of governance structures, such as river basin commissions and intergovernmental frameworks, has become central to dispute resolution and coordinated development. Third, the integration of scientific knowledge—including hydrological modeling, climate projections, and ecological assessment—has improved decision-making and facilitated equitable allocation (Gleick 1993). Finally, international legal principles, such as those codified in the UN Watercourses Convention (1997), provide normative guidance for cooperation, emphasizing equitable and reasonable utilization, the prevention of significant harm, and the obligation to notify and consult other riparian states before implementing measures that may affect shared waters (McCaffrey 2007).

To complement qualitative case narratives, this study incorporates a quantitative, spatial, and temporal examination of global transboundary water interactions using the Transboundary Freshwater Diplomacy Database (TFDD), developed by the *College of Earth, Ocean, and Atmospheric Sciences, Oregon State University* (Oregon State University 2025). The TFDD's Basins at Risk (BAR) Water Event Database systematically classifies more than 1800 international water-related events (1948–2008), each assigned an interaction intensity from −7 (most conflictual) to +7 (most cooperative) (Yoffe et al. 2003). This index transforms political and diplomatic exchanges into a measurable continuum, enabling consistent cross-basin comparison of cooperative and conflictive tendencies. The mean BAR Scale (Fig. 2.2) across transboundary basins offers a global snapshot of hydropolitical relationships. Red shades correspond to negative BAR values—indicating political, economic, or military conflict—whereas blue shades represent positive BAR values —indicating varying degrees of cooperation. Marker size reflects the magnitude of the mean BAR value (|BAR|), emphasizing locations where interaction intensity—whether cooperative or conflictual—is high. Each labeled basin corresponds to an official Basin Code (BCode) from the TFDD (e.g., *IND* = Indus, *NIL* = Nile, *MEK* = Mekong, *JDN* = Jordan, *DAN* = Danube, *TIG* = Tigris–Euphrates, *VOL* = Volta, *SEN* = Senegal). Overlapping labels were filtered to maintain visual clarity.

Spatially, apparent regional clustering is evident. Persistent conflict zones (−5 to −7) are concentrated in the Middle East, South Asia, and Horn of Africa, regions characterized by rapid development, climatic variability, and asymmetric power relations. By contrast, high cooperation levels (+5 to +7) are concentrated in Europe and North America, where long-standing treaties—such as the *Danube River*

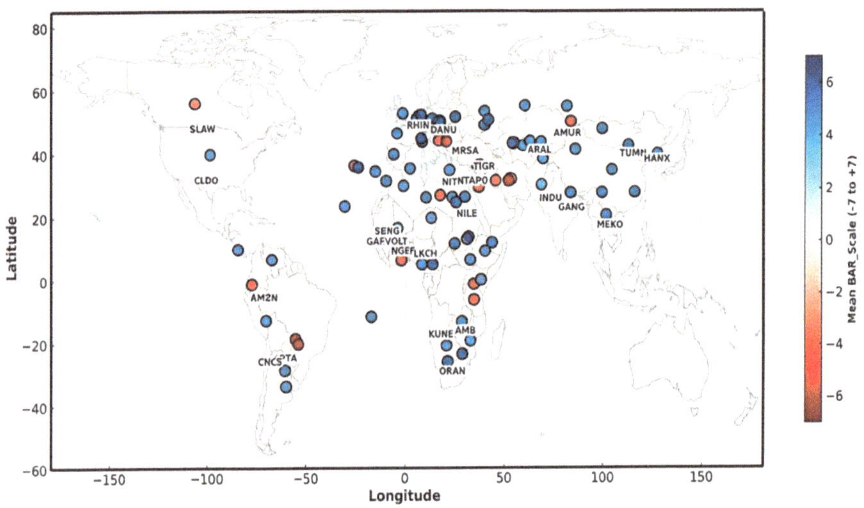

Fig. 2.2 Global distribution of mean BAR Scale values across transboundary river basins (1948–2008). The map illustrates global patterns of cooperation and conflict intensity based on mean BAR Scale values (−7 to +7) from the Transboundary Freshwater Diplomacy Database. Red shades denote conflictual events (−7 to −1), while blue shades denote cooperative interactions (+1 to +7); neutral or insignificant interactions (0) are shown as white or light tones. Marker size is proportional to the absolute magnitude of the mean BAR score for each basin-location pair, with larger markers reflecting stronger event intensity. Basin codes (BCode) correspond to the international river basins listed in the TFDD—e.g., IND: Indus, NIL: Nile, MEK: Mekong, JDN: Jordan, DAN: Danube, TIG: Tigris–Euphrates, VOL: Volta, and SEN: Senegal. Overlapping labels have been removed to preserve clarity. Data source: Transboundary Freshwater Diplomacy Database (Oregon State University, 2025); Yoffe et al. (2003)

Protection Convention and the *U.S.–Mexico 1944 Water Treaty*—provide durable institutional frameworks for collaboration. Transitional basins (−2 to +2), such as the Nile and Mekong, exhibit both conflictual and cooperative phases, reflecting the dynamic interplay between hydrological stress, economic development, and political negotiation.

Temporal analysis (Fig. 2.3) reveals how cooperation and conflict have evolved over six decades. Global mean BAR values show increasing cooperation after the early 1990s, corresponding with the end of the Cold War and a surge in treaty formation. Peaks of positive intensity coincide with significant institutional developments—the 1995 Mekong Agreement, the 1999 Nile Basin Initiative, and renewed Danube cooperation. Conversely, short-lived dips toward negative values align with episodes of geopolitical instability, such as Middle Eastern conflicts and South Asian disputes over dam construction. The 5-year smoothed trend confirms an overall shift toward cooperative engagements and institutional strengthening within transboundary water governance.

Beyond mapping and temporal trajectories, this analysis underscores the interdependence of environmental variability and political cooperation. Periods of sustained positive BAR values align with phases of enhanced global hydro-diplomacy, when scientific data exchange, joint monitoring, and climate-adaptation programs became institutional norms. Conversely, negative spikes often coincide with drought periods, upstream hydraulic nationalism, or the absence of multilateral frameworks. Thus, while hydro-political tensions are cyclical, their recurrence is mitigated where institutional flexibility, data transparency, and dispute-resolution mechanisms are embedded in basin governance. These results affirm that technical cooperation is not merely an outcome of political stability. Still, a catalyst for it, revealing that basins with adaptive

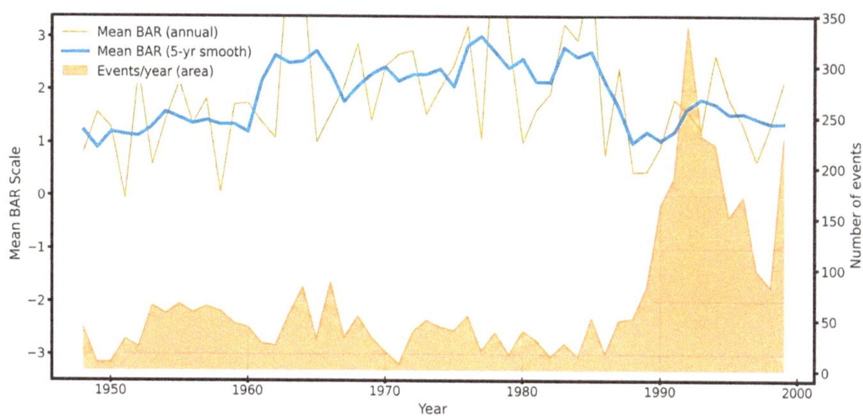

Fig. 2.3 Temporal trends in transboundary water-event intensity (1948–2008). Annual mean BAR Scale values (solid line) and 5-year smoothed averages (bold line) with event frequency (dashed line). Positive BAR values indicate cooperative behavior; negative values signify conflictual interactions. The post-1990 period shows a distinct cooperative trend, reflected in increased treaty activity and the development of basin institutions. Source: Oregon State University (2025); Yoffe et al. (2003)

and transparent governance—such as the Danube, Rhine, and Mekong—serve as empirical models for conflict transformation through sustained institutionalization.

2.3 Global Laws and Principles in Transboundary Water Management

The management of transboundary water resources is not solely a technical or administrative challenge; it is also a legal and normative one. Shared waters require governance structures that ensure equitable use, prevent harm, and promote cooperation among riparian states. Over the past century, international law has played a critical role in shaping principles and mechanisms for transboundary water management, providing normative guidance for states and establishing frameworks to prevent and resolve conflicts.

2.3.1 *Evolution of International Legal Frameworks*

The evolution of international water law reflects the growing recognition of water as a shared resource and a potential source of conflict. Early attempts at codification, such as the Helsinki Rules on the Uses of the Waters of International Rivers (1966), emphasized equitable and reasonable utilization of transboundary waters, recognizing both upstream and downstream interests. The Berlin Rules on Water Resources (2004) built upon this foundation by integrating sustainability, environmental protection, and human rights into the framework of water governance (McCaffrey 2007). The 1997 United Nations Convention on the Law of the Non-Navigational Uses of International Watercourses (UN Watercourses Convention) represents a milestone in global water law. The Convention establishes legally binding principles, including equitable and reasonable utilization, the obligation not to cause significant harm, the duty to cooperate, and procedures for notification and consultation regarding planned measures affecting shared watercourses (UN 1997). Although not yet universally ratified, the UN Watercourses Convention has influenced regional agreements and serves as a reference for best practices in transboundary water governance.

2.3.2 *Core Principles of Transboundary Water Law*

Several fundamental principles underpin the international legal framework for shared water management. Equitable utilization ensures that all riparian states can access and benefit from shared water resources, taking into account hydrological, social, economic, and environmental factors. This principle balances upstream and downstream interests, emphasizing fairness rather than absolute equality in allocation (McCaffrey 2007). States are legally obligated to prevent actions that may significantly harm other riparian states, whether through pollution, diversion, or excessive extraction. This principle encourages preventive action and environmental stewardship while promoting negotiation in the event of potential harm (Tanzi and Arcari 2001). Cooperation is central to transboundary water governance. The duty to consult, exchange information, and notify co-riparians of planned measures fosters transparency and trust. Mechanisms such as joint monitoring committees and river basin organizations operationalize these obligations, reducing the likelihood of disputes (Earle and Bazilli 2008). Modern legal instruments emphasize the ecological integrity of shared water systems. Sustainable management principles require states to maintain water quality, preserve biodiversity, and consider long-term ecological resilience in planning and utilization. This principle integrates environmental sustainability into the core of legal governance frameworks (Wouters 2006). Table 2.1 highlights the key global conflicts and resolution mechanisms.

2.3.3 Regional Agreements and Institutional Mechanisms

While global treaties provide broad normative guidance, regional agreements are essential for operationalizing these principles by adapting them to unique local hydrological, political, and socio-economic contexts. A primary example is the Indus Waters Treaty (1960), a landmark bilateral agreement between India and Pakistan that establishes specific water allocations and joint institutions for monitoring and conflict resolution (Kerr 2012). In a similar multilateral effort, the Nile Basin Initiative (1999) fosters cooperation among riparian states through joint management, data sharing, and sustainable development planning (Swain 2011). Regional bodies also play a critical role in ecological and resource stability across other continents. The Mekong River Commission (1995) facilitates essential dialogue and coordination among Southeast Asian countries, prioritizing sustainable development and rigorous ecological monitoring (Dore 2003). Furthermore, the SADC Protocol on Shared Watercourses (2000) provides a comprehensive legal framework for Southern Africa, establishing regional principles for equitable utilization, sustainable management, and institutional cooperation across multiple shared basins (Earle and Bazilli 2008).

These regional frameworks demonstrate how global legal principles are translated into actionable governance mechanisms, providing context-specific solutions while adhering to overarching norms. Despite comprehensive legal and institutional frameworks, challenges persist. Ratification of global conventions is uneven, enforcement mechanisms are often weak, and divergent national priorities can limit the effectiveness of agreements. Political instability, power asymmetry among riparian states, and lack of technical capacity further exacerbate these issues. Moreover, emerging challenges such as climate change, population growth, and

Table 2.1 Comprehensive summary of global transboundary water conflicts and resolution mechanisms

Conflict/basin	Countries/region	History of dispute	Governing principle/treaty	Game-theoretic application	Potency of resolution method	References
Indus river	India, Pakistan	1947–1960: Partition led to allocation disputes; 1960 Indus Waters Treaty signed	Indus Waters Treaty (1960); equitable utilization; dispute resolution mechanisms	Cooperative game theory: Nash bargaining to optimize water sharing; non-cooperative modeling of upstream dam use	High—Treaty has prevented major conflict; game theory supports joint infrastructure planning	Kerr (2012); Qureshi and Ahmad (2008)
Nile River	Egypt, Sudan, Ethiopia	1959–present: Upstream dam construction (GERD) vs. downstream water rights	UN Watercourses Convention (1997); Cooperative Nile Basin Initiative (1999)	Cooperative bargaining models for dam operation, revenue sharing, and non-cooperative risk modeling	Medium–High—Negotiations ongoing; models support Pareto-optimal outcomes	Seleshi et al. (2018); Swain (2011)
Tigris-Euphrates	Turkey, Syria, Iraq	1970s–present: Turkey's GAP dams reduce flow to Syria/Iraq	Helsinki Rules (1966); UN Watercourses Convention (1997)	Non-cooperative upstream extraction modeling; cooperative flow-sharing scenarios	Medium—Limited formal treaties; game-theoretic models suggest potential for cooperative solutions	Yusuf and Gupta (2004); Dinar and Wolf (1994)
Mekong River	Cambodia, Laos, Thailand, Vietnam	1990s–present: Hydropower development upstream vs. downstream impacts	Mekong River Commission (1995); UN Watercourses Convention	Cooperative games for dam scheduling; risk-sharing models for seasonal variability	Medium—Institutional cooperation exists, but enforcement is weak	Dore (2003); Earle and Bazilli (2008)
Jordan River	Israel, Jordan, Palestine, Syria	1948–present: Water allocation post-independence; political conflicts	Johnston Plan (1955); Peace treaties incorporating water allocation; UN Watercourses Convention	Bargaining games to allocate limited water; cooperative allocation under shared infrastructure	Medium–High—Political issues complicate enforcement; game theory supports fair allocation	Dinar and Wolf (1994); Tanzi and Arcari (2001)

Conflict/basin	Countries/region	History of dispute	Governing principle/treaty	Game-theoretic application	Potency of resolution method	References
Colorado River	USA, Mexico	1922–present: Over-allocation, droughts; treaty signed 1944	1944 US-Mexico Treaty; equitable allocation principles; modern agreements for surplus sharing	Cooperative games for water allocation under drought: simulation of usage and deficit sharing	High—Treaty provides a clear allocation; adaptive game-theory models improve flexibility	Wolf (2007); Dinar et al. (2007)
Danube River	Austria, Germany, Hungary, Romania, others	Nineteenth–twentieth century: Flood control, pollution; 1994 Danube River Protection Convention	Helsinki Rules; Danube River Protection Convention (1994)	Cooperative games for pollution reduction; joint infrastructure investment modeling	Highly effective multinational cooperation via EU frameworks	Wouters (2006); Earle and Bazilli (2008)
Ganges-Brahmaputra-Meghna	India, Bangladesh, Nepal, Bhutan	1950s–present: Flooding, irrigation disputes, hydroelectric projects	Ganges Water Treaty (1996 India-Bangladesh); UN Watercourses Convention; equitable utilization	Cooperative bargaining models; joint water-sharing simulations; cooperative infrastructure planning	Medium—Treaty mitigates downstream conflicts, but climate change and population pressures increase risk	Gupta et al. (2012); Qureshi (2013); Panday et al. (2025)
Krishna River	India (Maharashtra, Karnataka, Andhra Pradesh, Telangana)	1960s–present: Interstate water allocation disputes over irrigation and hydropower	Krishna Water Disputes Tribunal (1969, 2010); equitable utilization principles; Interstate Water Disputes Act	Cooperative bargaining for interstate sharing; non-cooperative modeling of reservoir operation	Medium—Tribunal rulings enforce allocations; game theory can optimize reservoir operation	Pandey et al. (2017)
Cauvery River	India (Karnataka, Tamil Nadu, Kerala, Puducherry)	Nineteenth century–present: Historical agreements and modern interstate disputes; Cauvery Water Disputes Tribunal (1990, 2007)	Cauvery Water Disputes Tribunal; Interstate Water Disputes Act; equitable allocation principles	Cooperative allocation models; Nash bargaining for seasonal water sharing	Medium—Tribunal and Supreme Court rulings are partially effective; game theory can improve seasonal water allocation	Ravilmar et al. (2015)

(continued)

Table 2.1 (continued)

Conflict/basin	Countries/region	History of dispute	Governing principle/treaty	Game-theoretic application	Potency of resolution method	References
Zambezi River	Zambia, Zimbabwe, Mozambique, Angola, Botswana	Twentieth century: Hydropower development, irrigation, and navigation conflicts	Zambezi Watercourse Commission (ZAMCOM, 2004); Helsinki Rules; UN Watercourses Convention	Cooperative game modeling for hydropower revenue sharing; risk-based water allocation models	Medium—Institutional cooperation exists, but enforcement varies; game theory can optimize shared use	Earle and Bazilli (2008); McCaffrey (2007)
Limpopo River	South Africa, Zimbabwe, Mozambique, Botswana	1980s–present: Flooding, water scarcity, irrigation	Limpopo Watercourse Commission (LIMCOM, 2003); UN Watercourses Convention	Cooperative and non-cooperative modeling of seasonal water allocation; cost-sharing for infrastructure	Medium—Institutional frameworks exist; game-theoretic analysis can guide equitable allocation	McCaffrey (2007); Earle and Bazilli (2008)
Volta River	Ghana, Burkina Faso, Togo, Côte d'Ivoire	1970s–present: Hydropower, irrigation, ecosystem management	Volta Basin Authority (VBA, 2007); UN Watercourses Convention	Cooperative allocation games: simulation of hydropower and irrigation trade-offs	Medium—Institutional capacity limited; game theory can optimize resource allocation	Earle and Bazilli (2008)
Orange River	South Africa, Namibia, Botswana, Lesotho	Twentieth century: Hydropower, irrigation, drought allocation	Orange-Senqu River Commission (ORASECOM, 2000); equitable utilization	Cooperative game modeling of water allocation: simulation under drought conditions	High—Strong institutional mechanism and treaty support; game theory reinforces equitable use	Turton (2003); Earle and Bazilli (2008)
Rhine River	Switzerland, Germany, France, the Netherlands	Nineteenth–twentieth century: Industrial pollution, flood control	International Commission for the Protection of the Rhine (ICPR, 1950s–1990s); Helsinki Rules	Cooperative games for pollution abatement and infrastructure planning	High—Effective multinational cooperation; successful water quality improvements	Wouters (2006); Earle and Bazilli (2008)

Conflict/basin	Countries/region	History of dispute	Governing principle/ treaty	Game-theoretic application	Potency of resolution method	References
Senegal River	Senegal, Mauritania, Mali, Guinea	1960s–present: Hydropower and irrigation disputes	Senegal River Basin Development Authority (OMVS, 1972); UN Watercourses Convention	Cooperative water allocation games: cost/ revenue sharing for dam management	Medium–High—Institutional cooperation is effective; game theory can optimize multi-country usage	Earle and Bazilli (2008); McCaffrey (2007)
Amu Darya/ Syr Darya	Uzbekistan, Kazakhstan, Tajikistan, Kyrgyzstan	1960s–present: Soviet-era irrigation and post-Soviet disputes; downstream water scarcity	Interstate Commission for Water Coordination (ICWC); UN Watercourses Convention	Cooperative and non-cooperative game models for irrigation scheduling; risk-sharing strategies	Medium—Institutional frameworks exist, but compliance challenges remain	Rasul and Sharma (2016); Dinar and Wolf (1994)

increased water demand require adaptive governance strategies that can respond to dynamic environmental and socio-economic conditions (Dinar et al. 2007).

2.4 Game-Theoretic Approaches in Transboundary Water Conflict Resolution

Transboundary water conflicts often arise from the competing interests of riparian states over limited water resources. Traditional legal and institutional mechanisms, while essential, may not fully capture states' strategic behavior when negotiating allocations, infrastructure development, or environmental management. Game theory, a branch of mathematics and economics that studies strategic interactions among rational decision-makers, provides a robust framework for modeling these conflicts, predicting outcomes, and designing mechanisms for cooperation. Game theory analyzes situations in which each player's payoff depends not only on their own actions but also on others' actions. In the context of transboundary water management, riparian states are the players, and water allocation decisions constitute the strategies. In non-cooperative games, each state acts independently to maximize its own utility, often leading to suboptimal outcomes known as the "tragedy of the commons" (Panday et al. 2021a, b). For instance, upstream states may over-extract water for irrigation or hydropower, harming downstream states. Non-cooperative frameworks highlight the potential for conflict and inefficiency but also clarify the incentives driving each actor (Dinar and Wolf 1994).

Cooperative game theory examines scenarios in which players can negotiate binding agreements to improve collective outcomes. In transboundary water management, this can involve water-sharing treaties, joint infrastructure projects, or coordinated environmental protection measures. Cooperative solutions often yield Pareto improvements, in which all states are better off than under unilateral action (Dinar et al. 2007). Bargaining models simulate negotiations over water allocation, cost-sharing, or investment in joint projects. These models account for power asymmetries, dispute resolution costs, and the strategic use of threats or promises. The Nash bargaining solution and its variants are commonly applied to identify fair allocations that balance equity and efficiency among riparian states (Dinar et al. 2007).

Game-theoretic modelling of the Nile Basin has analysed interactions between Egypt, Sudan, and Ethiopia regarding the Grand Ethiopian Renaissance Dam (GERD). Models of cooperative bargaining suggest that negotiated agreements on flow-release schedules and joint hydropower revenue sharing can yield Pareto-efficient outcomes, thereby reducing the risk of conflict (Seleshi et al. 2018). The Indus Waters Treaty provides a historical framework for water sharing between India and Pakistan. Game-theoretic simulations of treaty compliance and potential renegotiation scenarios indicate that cooperation is a dominant strategy for both parties, as unilateral defection would result in mutually harmful outcomes, including economic losses and increased political tension (Qureshi and Ahmad 2008). In the Tigris-Euphrates system, game-theoretic models have been used to study water allocation conflicts among Turkey, Syria, and Iraq. Non-cooperative models highlight the potential for upstream overuse, while cooperative models demonstrate that joint infrastructure planning and water banking strategies can optimize total basin benefits while protecting downstream needs (Yusuf and Gupta 2004).

When we look at the strategic landscape of managing transboundary water, game-theoretic methods go well beyond mere mathematical formulas; they provide a sophisticated lens for understanding the human and political frictions inherent in shared resources. One of the most significant advantages of this approach is its predictive power. Much like a chess player anticipating an opponent's next several moves, these models allow negotiators to look ahead and simulate how different states might react to specific policy changes. This foresight is crucial for effective planning and diplomacy. Furthermore, game theory excels at finding the "cooperative sweet spot." While traditional legal frameworks often focus on rigid rights and boundaries, cooperative games can uncover hidden, mutually beneficial agreements—actual "win-win" scenarios that prioritize shared prosperity over zero-sum competition. By quantifying complex trade-offs, these methods also bring clarity to the difficult choices between agricultural needs, infrastructure investments, and environmental protection. Ultimately, they serve as a vital tool for scenario analysis, helping leaders stress-test their strategies against the unpredictable pressures of climate change and shifting political tides.

However, translating these elegant mathematical solutions into real-world river basin management is a messy and challenging process. The most immediate hurdle is a heavy dependence on high-quality data. In the real world, reliable hydrological and economic figures are often scarce, or worse, treated as sensitive secrets by states that view data as a bargaining chip. There is also the "rationality trap." While game theory assumes every actor will behave logically to maximize their own utility, human history tells a different story. Real-world decisions are frequently swayed by domestic political pressures, long-standing historical grievances, or ideological stances that defy simple logic. As the number of countries involved in a basin grows, complexity can become overwhelming, often forcing researchers to oversimplify the intricate dynamics they seek to model. Perhaps most importantly, a mathematical "optimum" is only a starting point. For these solutions to actually result in flowing water and functioning dams, they must be carefully woven into our existing legal treaties and institutional mechanisms to ensure they are both implementable and legally sound.

2.5 Conclusion

This synthesis shows that transboundary water governance is shaped less by hydrological conditions alone than by the interaction of political power, institutional depth, and strategic behavior. Analysis of the TFPD and Basins at Risk data indicates that, although conflict persists in hydropolitically stressed regions, cooperative interactions have become increasingly prevalent, particularly in basins supported by comprehensive treaties, basin organizations, and data-sharing mechanisms. The post-1990 shift toward more positive BAR intensities confirms that cooperation is emerging as a dominant norm where institutional frameworks are mature and inclusive. Game-theoretic insights reinforce these findings by demonstrating that cooperation stabilizes when institutions realign incentives, reduce uncertainty, and credibly enforce agreements, while institutional gaps and unilateral actions sustain competitive equilibria. Looking forward, rising climate variability and demand pressures will test existing arrangements, underscoring the need for adaptive, transparent, and incentive-compatible governance. Proactive institutions that integrate data, flexibility, and strategic foresight are essential for transforming shared waters from sources of vulnerability into platforms for long-term cooperation and stability.

References

Allan JA (2001) The Middle East water question: hydropolitics and the global economy. I.B. Tauris

Dinar A (2003) Cooperative management of transboundary water resources: lessons from international experience. World Bank. https://doi.org/10.1596/0-8213-5384-7

Dinar A, Wolf AT (1994) Economic and political considerations in regional cooperation models. Agric Resour Econ Rev 23(1):44–59. https://doi.org/10.1017/S1068280500001159

Dinar S, Dinar A, McCaffrey S, McKinney D (2007) Bridges over water: understanding transboundary water conflict, negotiation and cooperation. World Scientific. https://doi.org/10.1142/9789812707084_0001

Dore J (2003) The governance of increasing Mekong regionalism. In: Kaosa-ard M, Dore J (eds) Social challenges for the Mekong region. Silkworm Books, pp 405–440

Earle A, Bazilli S (2008) A gendered critique of transboundary water management. Water Alternat 1(2):327–343

Gleick PH (1993) Water and conflict: freshwater resources and international security. Int Secur 18(1):79–112. https://doi.org/10.2307/2539033

Gupta J, van der Zaag P, Sarda R (2012) Institutional arrangements for river basin management: lessons from the Ganges-Brahmaputra-Meghna. Int J Water Resour Dev 28(3):475–491. https://doi.org/10.1080/07900627.2011.580421

Kerr M (2012) The Indus waters treaty: a history of successful negotiations. Water Int 37(7):871–882. https://doi.org/10.1080/02508060.2012.747982

Lowi MR (1993) Water and power: the politics of a scarce resource in the Jordan River basin. Cambridge University Press

McCaffrey SC (2007) The law of international watercourses: non-navigational uses. Oxford University Press

Mekong River Commission (1995) Agreement on the cooperation for the sustainable development of the Mekong River Basin. MRC, Vientiane. https://www.mrcmekong.org

Nile Basin Initiative (1999) Shared vision program: a framework for cooperative management. Nile Basin Secretariat. https://www.nilebasin.org

Oregon State University (2025) Transboundary Freshwater Diplomacy Database (TFDD). College of Earth, Ocean, and Atmospheric Sciences. https://transboundarywaters.oregonstate.edu

Panday DP, Khosa R, Maheswaran R et al (2021a) Game-theoretic-based modelling of the Krishna waters dispute: equilibrium solutions by Metagame Analysis. Eur Phys J B 94:101. https://doi.org/10.1140/epjb/s10051-021-00107-w

Panday DP, Khosa R, Maheswaran R et al (2021b) Game-theoretic modelling of Krishna water dispute equilibrium solutions via hypergame analysis. Eur Phys J B 94:131. https://doi.org/10.1140/epjb/s10051-021-00135-6
Panday DP, Kumari A, Kumar M (2025) Alkalinity-salinity-sustainability: decadal groundwater trends and its impact on agricultural water quality in the Indian Peninsula. Sci Total Environ 978:179459. https://doi.org/10.1016/j.scitotenv.2025.179459
Pandey R, Rao KG, Raju KS (2017) Application of game theory for water allocation in the Krishna River basin. Water Resour Manag 31(11):3559–3574. https://doi.org/10.1007/s11269-017-1677-5
Qureshi ME (2013) Water resource management in the Ganges basin: a comparison of policy and practice in Bangladesh, India, and Nepal. Water Policy 15(1):37–60. https://doi.org/10.2166/wp.2012.010
Qureshi AS, Ahmad Z (2008) Sustaining the Indus waters treaty: potential threats and measures to avert a crisis. Water Policy 10(3):203–217. https://doi.org/10.2166/wp.2008.041
Rasul G, Sharma B (2016) The nexus approach to water–energy–food security: An option for adaptation to climate change. Climate Policy 16(6):682–702
Ravilmar K, Natarajan S, Krishnan A (2015) Game-theoretic analysis of inter-state water disputes: case of the Cauvery River. J Hydrol 530:849–857. https://doi.org/10.1016/j.jhydrol.2015.10.017
Seleshi Y, Demissie SS, Ayana EK (2018) GERD, Nile Basin hydropolitics, and cooperative frameworks. Water Int 43(4):492–510. https://doi.org/10.1080/02508060.2018.1451124
Swain A (2011) Challenges for water sharing in the Nile Basin: changing geopolitical realities. Hydrol Sci J 56(4):687–702. https://doi.org/10.1080/02626667.2011.564175
Tanzi A, Arcari M (2001) The United Nations Convention on the Law of International Watercourses: a framework for sharing. Kluwer Law International
Turton AR (2003) Hydropolitics and the development of the South African economy. Afr J Confl Resolut 3(1):59–84
UN. (1997) Convention on the Law of the Non-Navigational Uses of International Watercourses. UN Treaty Ser 2999:77. https://treaties.un.org
UNESCO World Water Assessment Programme (2020) The United Nations world water development report 2020: Water and climate change. UNESCO
Wolf AT (1995) Hydropolitics along the Jordan river: the impact of scarce water resources on the Arab-Israeli conflict. United Nations University Press
Wolf AT (2007) Shared waters: conflict and cooperation. Annu Rev Environ Resour 32(1):241–269. https://doi.org/10.1146/annurev.energy.32.041006.101434
Wolf AT, Stahl K, Macomber MF (2003) Conflict and cooperation within international river basins: The importance of institutional capacity. Water Resources Update 125(1):31–40
Wouters P (2006) The legal response to international water conflicts: the UN Watercourses Convention and beyond. Water Resour Dev 22(4):555–565. https://doi.org/10.1080/07900620600770602
Yoffe S, Wolf AT, Giordano M (2003) Conflict and cooperation over international freshwater resources: indicators of basins at risk. J Am Water Resour Assoc 39(5):1109–1126. https://doi.org/10.1111/j.1752-1688.2003.tb03696.x
Yusuf SA, Gupta J (2004) The Tigris-Euphrates: a case study in transboundary water governance. Int J Water Resour Dev 20(2):209–225. https://doi.org/10.1080/0790062042000206158

Chapter 3
Innovative Water Conservation Practices: Managed Aquifer Recharge

Abstract Groundwater depletion has emerged as a critical global challenge due to increasing population, rapid urbanization, agricultural intensification, and climate variability. Managed Aquifer Recharge (MAR) offers a practical, cost-effective, and climate-resilient strategy to enhance water availability while simultaneously safeguarding groundwater quality. This chapter examines the principles, site considerations, and technologies associated with MAR, highlighting soil properties, water availability, climatic influences, and water quality as key determinants for successful implementation. Various MAR techniques, including surface spreading, recharge wells, bank filtration, in-channel modifications, and rainwater harvesting, are discussed with emphasis on their suitability under different hydrogeological and climatic conditions. Furthermore, the chapter highlights the importance of water quality management, particularly with respect to contaminants of emerging concern (CECs), including pharmaceuticals, pesticides, PFAS, and microplastics, which pose risks to groundwater integrity and human health. Global and national (India-specific) frameworks for monitoring and risk-based management are presented, along with future perspectives on MAR as a tool for achieving Sustainable Development Goals (SDGs) and enhancing climate adaptation. Overall, MAR is positioned as a cornerstone of sustainable water resource management, capable of mitigating water scarcity, ensuring groundwater security, and supporting ecosystem resilience.

Keywords Managed aquifer recharge · Water quality · SDGs · Groundwater · Innovative

3.1 Introduction

Groundwater depletion has emerged as a major driver of water scarcity worldwide. Groundwater supplies freshwater to approximately 1.5 billion people, yet increasing population growth and water demand are placing conventional water resources

M. Kumar, D. P. Panday, *Enabling Water Security and Sustainability: Global Challenges and Solutions*, SpringerBriefs in Water Science and Technology, https://doi.org/10.1007/978-3-032-21521-5_3

under severe stress (Alam et al. 2021). Irrigated agriculture, which accounts for nearly 70% of global freshwater use, relies heavily on groundwater in arid and semi-arid regions, particularly when surface water availability declines (Ingrao et al. 2023). Intensifying urbanization, competing sectoral demands, and climate extremes such as droughts and heatwaves have further exacerbated groundwater overexploitation, leading to stream depletion, land subsidence, seawater intrusion, aquifer storage loss, rising pumping costs, and groundwater quality degradation (Döll et al. 2012; Pachauri et al. 2014). Ensuring groundwater sustainability is therefore essential for meeting long-term water, food, and energy demands.

A global assessment of approximately 39 million groundwater wells across 40 countries highlights widespread declines in water table levels and increasing failure of shallow wells (Jasechko and Perrone 2021). In response, enhancing subsurface water storage has gained attention as a climate-resilient strategy, particularly through Managed Aquifer Recharge (MAR), which intentionally replenishes aquifers while safeguarding water quality for human and environmental health (Korkut et al. 2025). Unlike conventional artificial recharge, MAR integrates both quantity enhancement and quality management through pre- or post-treatment processes, ensuring suitability for intended uses (Sasidharan and Bradford 2025).

MAR is recognized for its cost-effectiveness, reduced evaporative losses compared to surface reservoirs, and flexibility in storing water from diverse sources, including rivers, stormwater, runoff, reclaimed wastewater, desalinated water, and inter-aquifer transfers (Dillon et al. 2019). Stored water can later be recovered for drinking, irrigation, industrial use, ecosystem support, and urban applications, while providing additional benefits such as controlling seawater intrusion, mitigating floods, improving groundwater quality, and preserving ecosystems. Historically, MAR has evolved from early irrigation practices (221 B.C.–1850) through industrial-era developments (1850–1950), post-war expansion and ASR innovation (1950–1990), to its current role as a globally adopted tool for addressing water scarcity, climate change, and environmental protection (Li et al. 2013; Zhang et al. 2020). Today, MAR is widely implemented in water-stressed regions, including the southwestern United States, northwestern India, Europe, and eastern Australia (Dillon et al. 2019; Sprenger et al. 2017; Panday et al. 2025; Kumar et al. 2025).

3.2 General Consideration on MAR

The selection of the MAR type depends on the particular requirements and site circumstances. The accessibility and quality of a water supply, regional hydrogeology, climate, groundwater demand, political environment, and social acceptability are key factors in determining the viability of a water supply (Alam et al. 2021). There are multiple MAR types, and the choice of method depends heavily on local site conditions and intended objectives. Key considerations include the availability and quality of source water, hydrogeological characteristics, climatic setting, groundwater demand, and political, regulatory, and sociocultural acceptance. From a technical standpoint, system design must be carefully aligned with the intervention's purpose (Ahirvar et al. 2025).

In many cases, MAR aims to enhance freshwater storage in aquifers for later use in domestic supply, agriculture, or industry, which requires the inclusion of an appropriate extraction mechanism. Beyond consumptive use, MAR also supports environmental functions such as improving groundwater quality, mitigating seawater intrusion, managing floods, safeguarding minimum river flows, and restoring groundwater-dependent ecosystems (Grinshpan et al. 2021; la Cecilia et al. 2024). The selection of MAR at a location is based on four main factors: (1) the soil's ability to absorb source water, (2) the need for water or scarcity, (3) the availability of source water influenced by climate, and (4) the requirements or issues related to water quality. Evaluating 1127 MAR projects (Stefan and Ansems 2017) and 33 research studies, we assess how these criteria influence the choice of MAR type at a location.

3.2.1 Soil Characteristics

Soil characteristics are a critical determinant in selecting suitable sites and methods for Managed Aquifer Recharge (MAR). Based on soil grain-size distribution, soils are commonly classified into four Hydrologic Soil Groups (HSGs): HSG-A, HSG-B, HSG-C, and HSG-D, broadly corresponding to sandy, sandy clay loam, clay loam, and clay-rich soils, respectively (Ross et al. 2018). Specifically, HSG-A soils contain >90% sand, HSG-B 50–90% sand and 10–20% clay, HSG-C <50% sand and 20–40% clay, and HSG-D <50% sand with >40% clay. Dual classes (e.g., A/D, B/D, C/D, and D/D) represent soils with high runoff potential unless artificially drained (Fig. 3.1). These classifications are widely used to rank MAR suitability in countries including the United States (O'Geen et al. 2015), India (Brindha and Pavelic 2016), Costa Rica (Bonilla Valverde et al. 2016), Jordan (Steinel et al. 2016), Spain (Stefan and Ansems 2017), and other regions (Fathi et al. 2020).

Globally, HSG-A soils are common in arid and semi-arid regions but often receive limited recharge due to climatic constraints, making climate a dominant factor in determining the need for MAR in such settings (Moeck et al. 2020). A global assessment of 1011 MAR sites by Stefan and Ansems (2017) showed that most MAR systems—excluding sand dams—are located in areas dominated by HSG-C soils. Hydrologic soil group D (HSG-D) is typically the second-most-preferred soil type for channel infiltration, spreading methods, and check dams, whereas HSG-B is favored for bank filtration and aquifer storage and recovery (ASR). While HSG classification provides a valuable first-order estimate of surface infiltration potential, subsurface permeability often differs due to geological heterogeneity. Consequently, a comprehensive evaluation of the subsurface geological framework is essential for accurately assessing MAR feasibility at any site.

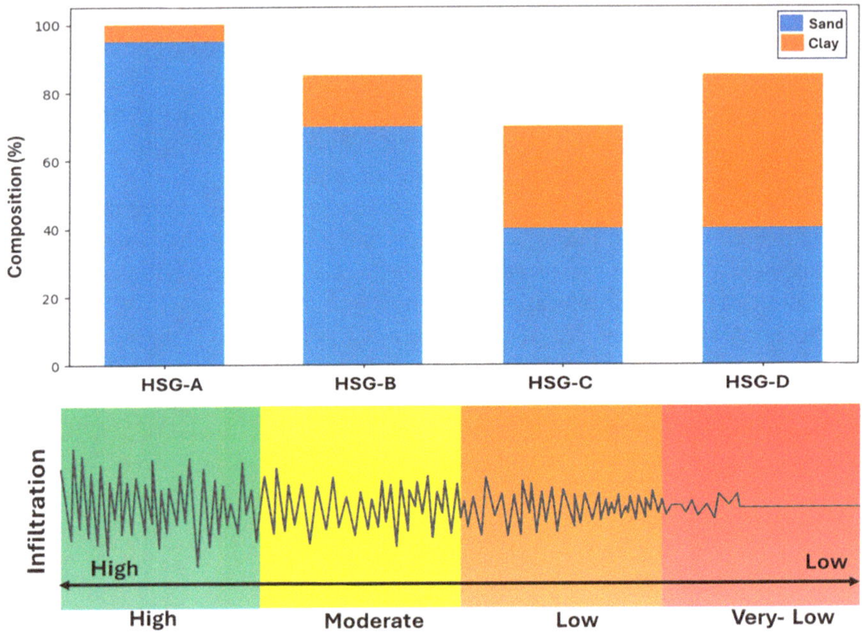

Fig. 3.1 Hydrologic Soil Groups (HSG) classification (**A, B, C, D**) based on grain-size distribution and infiltration potential, relevant to MAR site selection

3.2.2 Trends in Water Availability and Long-Term Water Storage Trend

The spatial distribution of Managed Aquifer Recharge (MAR) sites relative to water availability has been evaluated at the basin scale, where availability is defined as the long-term balance between precipitation (P) and evapotranspiration (ET) (Alam et al. 2021). Results indicate that MAR systems designed primarily for groundwater storage are typically implemented in regions with low water availability, whereas MAR projects focused on water quality improvement are more commonly located in water-rich areas. Certain MAR types, such as soil–aquifer treatment (SAT), are prevalent in water-stressed urban environments, where recycled water supplements limited natural supplies.

To identify regions where MAR could provide the most significant benefit under high demand and limited availability, long-term trends in terrestrial water storage (TWS) from 2002 to 2017 were analyzed using GRACE satellite data ($p < 0.05$). The analysis revealed widespread global declines in TWS, signaling an increased risk of future water scarcity. Water availability (P–ET) at MAR sites with water quality objectives was significantly higher ($p < 0.00001$) than at storage-oriented MAR sites, indicating that storage-oriented. MAR is more common in areas experiencing declining water availability. In such regions, MAR systems that use alternative sources, such as stormwater and recycled water, are becoming increasingly popular. Consequently, MAR site selection should prioritize areas with high water demand and seasonal water availability suitable for infiltration and subsurface storage.

Local climate strongly influences both water availability and MAR system selection. In arid and semi-arid regions, agriculture depends heavily on groundwater, especially during droughts when excessive abstraction of saline groundwater can degrade soils (Faunt et al. 2016). MAR mitigates these impacts by diluting salinity and limiting seawater intrusion. Climate change is projected to intensify droughts and extreme precipitation events, increasing runoff losses and water supply uncertainty despite episodic recharge opportunities, thereby threatening food security (Cuthbert et al. 2019). MAR is therefore particularly valuable in drought-prone, groundwater-dependent regions, where it sustains dry-season flows, enhances ecological conditions, and supports riverine ecosystems, especially in Mediterranean climates with peak demand during periods of surface-water scarcity (Mukherjee et al. 2018).

Across the five major climate classes—humid, dry sub-humid, semi-arid, arid, and hyper-arid—surface water from rivers and lakes remains the dominant source for MAR, followed by stormwater and increasingly by recycled water (Fig. 3.2). In semi-arid, dry sub-humid, and humid regions, contaminated groundwater is also commonly used, particularly in subsurface and sand dam systems that create underground storage by impeding groundwater flow. In hyper-arid regions, limited rainfall renders recharge from precipitation economically unfeasible, making treated

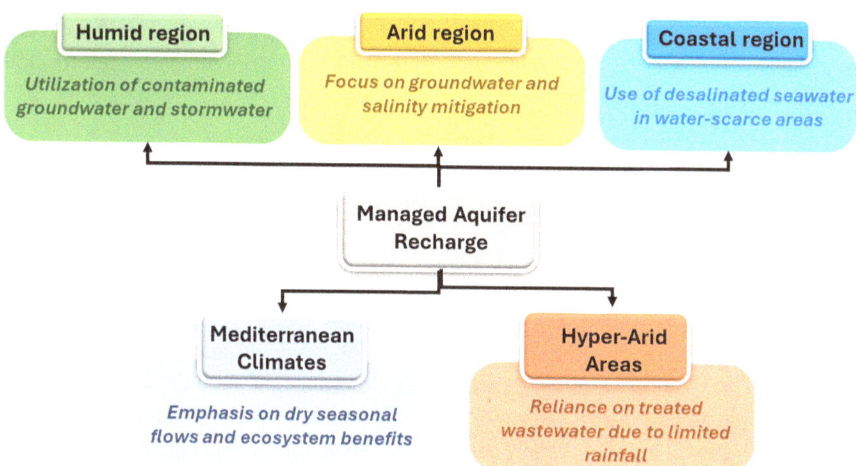

Fig. 3.2 Regional applications of Managed Aquifer Recharge across different climatic settings, including the use of contaminated groundwater and stormwater in humid regions, salinity mitigation in arid areas, desalinated seawater in coastal regions, seasonal flow support in Mediterranean climates, and reliance on treated wastewater in hyper-arid regions

wastewater a more reliable MAR source. In coastal settings, desalinated seawater plays a critical role, especially in Middle Eastern countries such as Israel and Kuwait, where alternative freshwater resources are minimal (Ganot et al. 2018; Al-Awadi et al. 1995).

3.2.3 Selection of MAR Based on Water Quality

The quality of source water and the ability of a particular MAR method to enhance water quality dictate whether MAR can be applied at a location. For example, MAR can be utilized in areas where the discharge of stormwater and wastewater into surface waters raises ecological issues (Masoner et al. 2019; Page et al. 2018). Similarly, agricultural runoff contaminates surface waters. In such contexts, MAR offers a sustainable option for both the safe disposal and treatment of contaminated water. Beyond mitigating surface water pollution, the use of treated water also minimizes the reliance on long-distance conveyance through canals or dams.

While MAR is primarily used to remove pollutants from source water and recharge aquifers, it can also dilute contaminated groundwater. The effectiveness of pollutant removal depends on the contaminant type, and poor implementation may introduce trace amounts of organics, pesticides, pathogens, nutrients, or disinfection by-products into groundwater. To support safe application, pollutants are categorized into low-, medium-, and high-risk groups.

Natural soil processes can effectively remove low-risk pollutants and require no pretreatment. Medium-risk pollutants generally necessitate pretreatment, such as sand filtration or UV disinfection. High-risk, recalcitrant contaminants, including PFAS, dioxanes, and NAPLs, are unlikely to be removed and pose significant risks to groundwater (Adler et al. 2025). Accordingly, MAR design must account for both source water quality and aquifer redox conditions.

3.3 Technologies and Methods of MAR

The last few decades have witnessed record groundwater pumping, the development of new technologies for water overdraft, and advancements in treatment, collectively driving progress in intentional groundwater replenishment or managed aquifer recharge (Dillon et al. 2018). A variety of structures and techniques can enhance aquifer water recharging and storage. The selection of a suitable method for sustainably recharging an aquifer system depends on the aquifer's hydrogeological characteristics, the source of the water, and the intended use of the water after it's recovered (Sufyan et al. 2024).

MAR techniques are broadly grouped into surface infiltration methods and well-based recharge methods, depending on the depth and mechanism of water entry into the subsurface (Dillon et al. 2019). Surface methods are best suited to unconfined aquifers with permeable layers and are classified by their location relative to natural watercourses. In-channel systems, such as check dams and sand dams, are constructed within riverbeds or streams to promote infiltration of flowing water. Off-channel systems, on the other hand, include recharge ponds, flood-irrigated fields, infiltration canals, trenches, or furrows, positioned outside the main watercourse. Where surface infiltration is not feasible, such as in confined aquifers or areas with limited permeable surfaces, deep recharge methods using injection wells deliver water directly into the aquifer (Alam et al. 2021). MAR technologies are classified based on recharge and storage technique into five broad categories as follows: (1) mechanisms of spreading, (2) in-channel modifications, (3) well or recharge shaft and boreholes, (4) induced bank filtration, and (5) runoff harvesting, as shown in Table 3.1.

3.3.1 Surface Spreading Techniques

Spreading methods are the most widely used and cost-effective approaches for Managed Aquifer Recharge (MAR). These include infiltration ponds and basins, agricultural MAR (Ag-MAR), check dams, bench terracing, soil–aquifer treatment (SAT), controlled flooding, and excess irrigation (Asano and Cotruvo 2004; Ghasemizade et al. 2019). Infiltration ponds or basins—either excavated or embanked—store water from sources such as stormwater, rainfall, or dam releases, allowing it to percolate through permeable soils and recharge underlying aquifers. By increasing water–soil contact time and surface area, these systems enhance infiltration efficiency and long-term groundwater storage in phreatic aquifers (Sufyan et al. 2024).

This category also includes dams, small reservoirs, stormwater treatment facilities, single-facility wastewater treatment plants, reservoir pavements, recharge pits, drainage and vegetated ditches, mound systems, sand filters, and drain fields (Casanova et al. 2016). However, spreading systems are susceptible to physical, chemical, and biological clogging processes, including particle rearrangement, clay deflocculation, mineral precipitation, ion exchange, scale formation, and microbial or algal growth (Table 3.1). Among spreading techniques, SAT systems are ubiquitous, using intermittent infiltration to recharge aquifers and improve water quality, mainly where unconfined aquifers occur near the surface (Zhang et al. 2020).

Spreading methods require unsaturated soil conditions maintained through periodic application and are most effective in permeable formations such as sandstone or alluvium (Mohammad-Hosseinpour and Molina 2022). As recharging water migrates through the unsaturated zone, natural attenuation processes reduce organic matter, trace metals, and microbial contaminants via physical, chemical, and biological mechanisms. Despite these advantages, the requirement for large, open land areas with suitable topography limits the applicability of spreading methods, particularly in densely urbanized settings.

Table 3.1 Summary of managed aquifer recharge (MAR) technologies, aquifer type, clogging mechanisms, application conditions, and prevention strategies

MAR types	MAR sub-type	Aquifer type	Clogging type	Clogging process	Prevention and rehabilitation	Application
Spreading methods	• Infiltration ponds and basins; • Soil aquifer treatment (SAT); • Excess irrigation, • Ditches and trenches, • Sprinkler irrigation	Alluvium Sandstone Carbonate aquifers	Physical, Chemical & Biological	**Physical:** Particle rearrangement from recharge–recovery cycles. Velocity-induced damage. Clay deflocculation Invasion of drilling fluid emulsifiers **Chemical:** Mineral precipitation (e.g., Fe, Al, $CaCO_3$ growth). Aquifer matrix dissolution (may increase conductivity). Ion exchange Ion adsorption Oxygen reduction Insoluble scale formation dissolution **Biological:** Algae growth & biological floc accumulation Microbial polysaccharide production Bacterial entrainment & growth	**Physical:** Water quality control (pretreatment); Backwashing. Backwashing: Proper recharge method. **Chemical**: Water quality control (pretreatment); Acidizing treatment; Backwashing. **Biological:** Water quality control (pretreatment); Sterilization; Backwashing; Well washing.	Unconfined aquifer at/near ground surface recharged via infiltration through permeable material.
Well, shaft and borehole recharge	• Open wells and shafts; • Aquifer storage and recovery (ASR); • Aquifer storage, transfer, and recovery (ASTR).	Deep and clay-covered aquifers				Occurs where an impermeable layer covers deep, clay-rich aquifers.
In-channel modifications	• Percolation ponds; • Sand storage dams; • Subsurface dams; • Leaky dams and • Recharge releases.	Alluvial Aquifers				Directs runoff for water retention and storage, especially during floods.
Induced bank infiltration	• Bank filtration; • Dune filtration.	Dry rivers with Dams or perennial rivers or streams with adjacent permeable sand layers.				Near a surface water body, lower pressure is required to induce water infiltration into the aquifer.
Run off/rainwater harvesting.	• Rainwater harvesting; • Rainwater recharge from open spaces.	Alluvium, sandstone, and carbonate aquifers				Where runoff can be collected for productive use.

3.3.2 *Recharge Wells and Borehole Injection*

In this class of MAR application, water is infiltrated through wells directly into the target aquifer. These techniques include open shafts and wells, aquifer storage and aquifer storage, transport, and recovery (ASTR) and recovery (ASR), which are utilized in most nations where deep aquifers are common. Injection methods are primarily used in medical settings (Dupont et al. 2018). These methods are generally used when the unsaturated zone restricts water infiltration, when the aquifer is overlain by a confining layer, or for reusing existing shallow wells.

In this approach, water is stored within the aquifer and later recovered either from the same injection well (ASR) or from a separate well (ASTR), the latter allowing for additional treatment through extended residence time in the aquifer (Sprenger et al. 2017). ASR refers to the injection of water into a favourable aquifer through a well during times of excess and the withdrawal of water from the same well during times of need (Pyne 2005). In contrast, ASTR involves aquifer flow between production and injection wells. ASTR is used to balance seasonal shortfalls in source water by injecting water during periods of excess and withdrawing it during periods of demand.

3.3.3 *Bank Filtration and Infiltration Galleries*

Induced bank filtration is an effective way to enhance the quantity and quality of water, particularly for poor-quality surface water sources such as rivers or lakes. Induced bank filtration involves constructing a series of parallel wells to a water body, and pumping creates conditions favorable for the infiltration of surface water through adjacent sediments into the aquifer (Arnaud et al. 2020). As an indirect recharge method, induced bank filtration encompasses both bank filtration and dune filtration techniques. Bank filtration involves extracting groundwater from wells near a surface water body, thereby lowering the groundwater table and creating a hydraulic gradient that forces surface water into the aquifer (Dillon 2005; Casanova et al. 2016). As seeping water travels through the porous sediments of the riverbank, it undergoes natural purification, significantly improving its quality.

Dune filtration operates on the same principle: surface water seeps through the dunes before being reclaimed by low-lying wells or ponds. This technique increases residence time in the subsurface, with additional physical, chemical, and biological treatment, and a travel time of more than 1 month is often required to achieve acceptable purification (Maliva and Missimer 2012). In contrast, infiltration galleries are buried trenches (containing polythene cells or slotted pipes) in permeable soils that allow infiltration through the unsaturated zone to an unconfined aquifer.

3.3.4 In-Channel Modifications

In-channel modification techniques for Managed Aquifer Recharge (MAR) involve altering the natural flow of rivers, streams, or canals to promote vertical infiltration and augment groundwater replenishment (Zhang et al. 2020). These interventions often employ engineered structures to regulate water movement and enhance the contact time between surface water and permeable streambed materials. In intermittent streams, recharge dams constructed across the streambed can intercept flows, enabling controlled downstream releases that align with the infiltration capacity of the underlying aquifer (Dupont et al. 2018).

Behind such dams, water retention promotes percolation, and in cases where streambeds are impermeable, upstream deposition of sand and gravel can create artificial aquifers for storing stormwater runoff. Where intermittent streams overlie shallow bedrock, subsurface barriers composed of low-permeability materials can be installed to retain runoff within alluvial deposits (Standen et al. 2020). In perennial streams, flow modifications, such as installing L-shaped levees, can expand infiltration zones and reduce flow velocity, thereby facilitating recharge. Common in-channel modification structures include percolation ponds, sand storage dams, subsurface dams (including leaky dams), and managed recharge releases, all designed to optimize stormwater capture and infiltration into adjacent or underlying aquifers.

3.3.5 Run off/ Rainwater Harvesting

Rainwater harvesting for MAR can be implemented at scales from individual households to entire villages, with the collected water directed to storage systems that replenish groundwater. A variety of structures, including trenches, reverse drainage systems, and storage tanks, are used to capture and convey rainwater. In urban areas, rooftop rainwater harvesting has gained prominence for both sustaining groundwater levels and mitigating stormwater runoff (Jakeman et al. 2016). In many MAR applications, harvested rainwater is routed to deep percolation pits, where it infiltrates the soil and recharges underlying aquifers, and can then be reused for various purposes. This method enhances the natural filtration of rainwater, improves aquifer replenishment, and contributes to restoring the water balance within the hydrological cycle, particularly in urban environments (Kim et al. 2007). Comprehensive descriptions of these MAR techniques are presented in Table 3.1, while Fig. 3.3 illustrates the diversity of MAR systems and their practical applications, and Fig. 3.4 highlights the evolution of MAR methods and future directions.

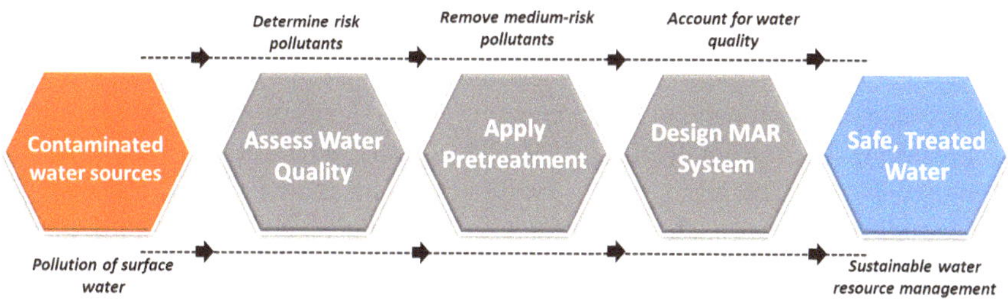

Fig. 3.3 Framework for water quality management in MAR, illustrating the process from contaminated water sources through quality assessment, pretreatment, and system design, leading to safe, treated water

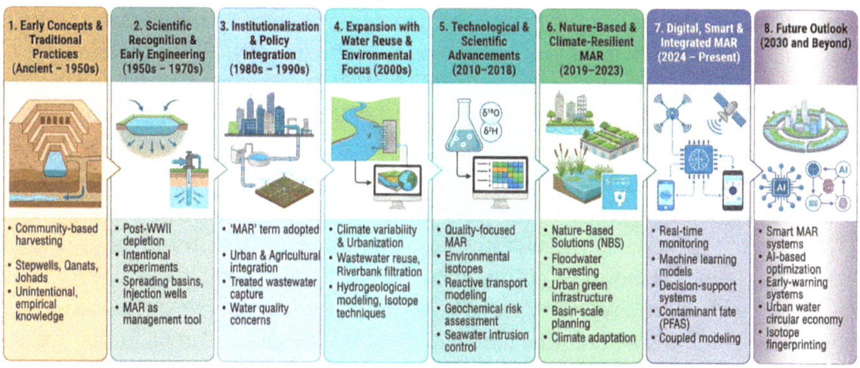

Fig. 3.4 Evolution of Managed Aquifer Recharge (MAR) from traditional water-harvesting practices to advanced, climate-resilient, and digitally enabled groundwater management strategies. The timeline highlights key scientific, technological, and policy milestones shaping modern MAR applications

3.4 Framework for Water Quality Management in MAR System

The water supplied through MAR systems is designed to be fit-for-purpose, implying that quality requirements vary according to the intended use, whether for mitigating salinization, irrigation, or potable supply. Consequently, water quality considerations are central to realizing the full potential of MAR. These span from improvements or possible deterioration in quality during subsurface passage to operational challenges such as well clogging, as well as broader sustainability concerns, including the reuse of treated wastewater effluent and stormwater (Kazner et al. 2012). Increasing anthropogenic pressures, heightened environmental awareness, and advancements in analytical capabilities have further intensified the focus on water quality within MAR applications.

3.4.1 India

Water quality management is fundamental to the safe and sustainable implementation of Managed Aquifer Recharge (MAR) in India, where MAR is increasingly promoted to enhance groundwater availability. The Indian framework adopts a risk-based, staged approach, beginning with preliminary screening to exclude low-risk activities, such as rooftop rainwater harvesting into shallow, unconfined aquifers not used for drinking. Projects with identified risks undergo a viability assessment evaluating source-water availability, aquifer suitability, storage capacity, and demand. For unconfined aquifers recharged with natural runoff, additional sanitary surveys assess potential pathogen and chemical hazards from sources such as latrines, septic systems, agriculture, or industry. In contrast, aquifer assessments evaluate risks related to clogging, geochemical mobilization of arsenic and fluoride, and proximity to groundwater users or sensitive ecosystems (Government of India, Ministry of Water Resources 2012). Risk management is guided by the Water Safety Plan (WSP) framework, which defines system mapping, control measures, operational limits, corrective actions, and verification procedures. Control measures include catchment protection (e.g., fencing, setback distances, sealed wells) and pretreatment options such as sedimentation basins and sand filters (WHO 2011). Where recharge water is intended for potable use, post-treatment measures—such as boiling, chlorination, ceramic filtration, or advanced treatment—may be required. Preventive actions, including livestock exclusion, improved sanitation, and routine maintenance of recharge structures, further reduce contamination risks.

3.4.2 Monitoring and Maintenance Strategies

Monitoring is central to MAR performance and safety. A core set of parameters—including pH, electrical conductivity, turbidity, nutrients, organic matter, major ions, fluoride, arsenic (where relevant), and E. coli—should be monitored at least quarterly in groundwater and three times per season in surface-water sources. Higher-risk projects, such as those involving sewage-derived recharge or aquifers linked to drinking-water supplies, require expanded monitoring and quantitative risk assessment. Effective governance strengthens water quality management by aligning MAR projects with state and national policies, clarifying source-water entitlements, and disseminating results from demonstration sites. While the Indian MAR Guide provides a practical, low-data framework for many applications, projects involving confined aquifers, urban stormwater, or industrial effluents must comply with more stringent international guidelines (Agarwal et al. 2024).

3.4.3 Global Framework

On a global scale, water quality remains the cornerstone of MAR implementation. Recharge water must meet fit-for-purpose standards, determined by its intended end use. While soil–aquifer treatment (SAT) and subsurface processes can improve water quality by attenuating microbial and chemical contaminants, these natural processes are highly site-specific. They may also lead to adverse reactions such as the mobilization of salts, trace metals, or organic matter. Consequently, effective water quality management is essential not only to protect aquifers but also to ensure the long-term acceptance and sustainability of MAR worldwide.

Internationally, regulatory frameworks for MAR are uneven. In countries such as the USA, Spain, Italy, Israel, and Australia, relatively advanced systems are in place, often with strict Maximum Allowable Concentrations (MACs) and detailed monitoring protocols. For instance, California and Israel apply some of the world's most stringent potable reuse standards, while Australia has pioneered risk-based guidelines that emphasize hazard assessment and multi-barrier safety. In contrast, many regions in Africa, Asia, and Latin America lack MAR-specific regulations and instead rely on general water reuse rules or the World Health Organization (WHO) guidelines. While the WHO framework provides a helpful baseline for safe practice, it is less stringent than the tailored standards adopted in advanced jurisdictions. This fragmented regulatory landscape underscores the need for greater harmonization of international MAR policies.

3.5 Ensuring Water Quality in MAR: Addressing Emerging Contaminants

Managed Aquifer Recharge (MAR) involves the intentional infiltration of surface water into aquifers to mitigate groundwater depletion and ensure adequate water volumes of suitable quality for drinking and other uses (Scanlon et al. 2023). However, surface waters commonly used for MAR often contain contaminants of emerging concern (CECs) at environmentally relevant concentrations due to anthropogenic inputs (Christensen et al. 2022). Consequently, MAR may serve as a pathway for CECs to enter groundwater systems and drinking-water supplies, posing potential risks to human health and ecosystems (Fakhreddine et al. 2021). Despite growing concern, the processes governing CEC transport and attenuation during MAR remain poorly understood.

Most CECs are anthropogenic, persistent, and widely distributed in the environment. Surface waters are particularly vulnerable as they drain large catchments contaminated by both point and diffuse sources, including inadequately treated wastewater, industrial discharges, agricultural runoff, and atmospheric deposition. Stormwater and roof runoff further contribute to CEC loads transported into recharge systems.

3.5.1 Types of CECs Detected in MAR Contexts

A broad range of CECs has been reported in MAR systems, including pharmaceuticals, personal care products (PPCPs), pesticides, industrial chemicals, endocrine-disrupting compounds (EDCs), and microplastics. Pharmaceuticals—such as antibiotics and anti-inflammatories—are frequently detected due to their widespread use and persistence in wastewater and surface waters. PPCPs, including fragrances, preservatives, and UV filters, are also common, reflecting domestic consumption patterns. Agricultural runoff introduces pesticides and herbicides, which are of particular concern due to their mobility and persistence in aquifers. EDCs (e.g., bisphenol A, phthalates, and hormones) and microplastics are increasingly reported, especially where treated wastewater or stormwater is used for recharge. The diversity of detected CECs across MAR systems highlights the need for site-specific monitoring and treatment strategies (Kazner et al. 2012; Page et al. 2018).

3.5.2 Risks to Groundwater Quality and Human Health

CECs in MAR systems pose risks due to their persistence, mobility, and incomplete removal during soil–aquifer treatment. Trace concentrations of pharmaceuticals, pesticides, and EDCs have been shown to persist in aquifers, raising concerns about long-term accumulation and mixture effects. Human health risks include endocrine disruption, carcinogenicity, neurotoxicity, and the proliferation of antibiotic resistance genes (ARGs) within groundwater microbial communities. Beyond health impacts, detecting CECs—even at low concentrations—can undermine public acceptance of MAR-derived water. These challenges underscore the need for advanced monitoring, risk-based frameworks, and precautionary management to protect groundwater quality and public health (Drewes et al. 2017; NRMMC-EPHC-NHMRC 2009).

3.5.3 Role of MAR in Sustainable Development Goals

MAR contributes directly to SDG 6 (Clean Water and Sanitation) by increasing groundwater availability and improving water quality through natural attenuation. Indirectly, it supports SDG 2 (Zero Hunger) by securing irrigation supplies, SDG 3 (Good Health and Well-being) by safeguarding drinking-water quality, SDG 13 (Climate Action) by increasing resilience to climate variability, and SDG 15 (Life on Land) by sustaining groundwater-dependent ecosystems (UN 2015; Villholth et al. 2016).

MAR is increasingly recognized as a climate adaptation strategy, buffering drought impacts by storing water during wet periods and mitigating flood risks by infiltrating stormwater. It also plays a key role in managing salinity and seawater intrusion in coastal aquifers threatened by sea-level rise and over-abstraction (Guppy et al. 2018). In arid and semi-arid regions, MAR enhances aquifer resilience under shifting rainfall regimes, stabilizing water supplies for communities, agriculture, and ecosystems. As climate extremes intensify, MAR will remain central to integrated adaptation strategies combining engineering solutions, nature-based approaches, and groundwater governance (Dillon et al. 2019).

3.6 Future Directions and Emerging Trends

Despite its proven benefits, several research gaps remain in understanding and optimizing MAR systems. One major challenge is the fate and transport of contaminants of emerging concern (CECs), such as pharmaceuticals, PFAS, and microplastics, which are increasingly detected in recharge waters but not yet fully regulated. There is also a need for advanced tools to predict better aquifer responses to a changing climate, particularly in data-scarce regions. Emerging trends highlight the integration of digital monitoring technologies, including real-time sensors and remote sensing, which enhance performance tracking and risk management. Coupling MAR with green infrastructure, such as wetlands, rain gardens, and permeable pavements, is gaining traction as a multifunctional strategy that integrates groundwater recharge with urban water management and ecosystem services. Furthermore, socio-political dimensions, including public acceptance, governance, and financing mechanisms, remain underexplored yet critical for scaling up MAR globally. Addressing these gaps will be essential to positioning MAR as a cornerstone of resilient, sustainable water management in the coming decades.

References

Adler O, Adeyemo T, Turkeltaub T, Gafni A, Weisbrod N, Bar-Zeev E (2025) Impact of redox potential on biofilms and water quality in managed aquifer recharge lab system. ACS ES&T Water

Agarwal V, Kumar M, Panday DP, Zang J, Munoz-Arriola F (2024) Unlocking the potential of remote sensing for arsenic contamination detection and management: challenges and perspectives. Curr Opin Environ Sci Health:100578. https://doi.org/10.1016/j.coesh.2024.100578

Ahirvar BP, Das P, Kumar R, Panday DP, Das N, Srivastva PK, Kumar M (2025) Weathering in the subsurface environment of the Gondwana deposits intensified by the cumulative impacts of mining and agriculture practices. J Contam Hydrol:104630. https://doi.org/10.1016/j.jconhyd.2025.104630. (IF: 4.4; Q1)

Alam S, Borthakur A, Ravi S, Gebremichael M, Mohanty SK (2021) Managed aquifer recharge implementation criteria to achieve water sustainability. Sci Total Environ 768:144992

Al-Awadi E, Mukhopadhyay A, Al-Haddad AJ (1995) Compatibility of desalinated water with the Dammam formation at the north-west Shigaya water-well field, Kuwait—a preliminary study. Hydrogeol J 3(4):56–73. https://doi.org/10.1007/s100400050068

Arnaud S, Staša B, Josip T, Boris B, Vik S, Božidar Č, Vedran S, Elisabeth M, Mbfsz A, Horváth M, Ferenc S, Mbfsz P, Selmeczi M, Tamás C, Lilian F, Magdolna A, István S, Antal S, Geogold K, Anne I (2020) Collection of good practices and benchmark analysis on MAR solutions in the EU

Asano T, Cotruvo JA (2004) Groundwater recharge with reclaimed municipal wastewater: health and regulatory considerations. Water Res 38(8):1941–1951

Bonilla Valverde JP, Blank C, Roidt M, Schneider L, Stefan C (2016) Application of a GIS multi-criteria decision analysis for the identification of intrinsic suitable sites in Costa Rica for the application of managed aquifer recharge (MAR) through spreading methods. Water 8(9):391. https://doi.org/10.3390/w8090391

Brindha K, Pavelic P (2016) Identifying priority watersheds to mitigate flood and drought impacts by novel conjunctive water use management. Environ Earth Sci 75(5):399. https://doi.org/10.1007/s12665-015-4989-z

Casanova J, Devau N, Pettenati M (2016) Managed aquifer recharge: an overview of issues and options. In: Integrated groundwater management: concepts, approaches and challenges. Springer International Publishing, pp 413–434

Christensen ER, Wang Y, Huo J, Li A (2022) Properties and fate and transport of persistent and mobile polar organic water pollutants: a review. J Environ Chem Eng 10(2):107201. https://doi.org/10.1016/j.jece.2022.107201

Cuthbert MO, Taylor RG, Favreau G, Todd MC, Shamsudduha M, Villholth KG et al (2019) Observed controls on resilience of groundwater to climate variability in sub-Saharan Africa. Nature 572(7768):230–234

Dillon P (2005) Future management of aquifer recharge. Hydrogeol J 13(1):313–316

Dillon P, Stuyfzand PJ, Grischek T, Liuria M et al (2018) Sixty years of global progress in managed aquifer recharge. Hydrogeol J. https://doi.org/10.1007/s10040-018-1841-z

Dillon P, Stuyfzand P, Grischek T, Lluria M, Pyne RDG, Jain RC, Bear J, Schwarz J, Wang W, Fernandez E (2019) Sixty years of global progress in managed aquifer recharge. Hydrogeol J 27(1):1–30

Döll P, Hoffmann-Dobrev H, Portmann FT, Siebert S, Eicker A, Rodell M et al (2012) Impact of water withdrawals from groundwater and surface water on continental water storage variations. J Geodyn 59:143–156

Dupont F, Sterckx A, Dijksma R (2018) Managed aquifer recharge (MAR) suitability maps and standardized suitability index, the case study of the Occitanie region (South France). International Groundwater Resources Assessment Centre, Francia

Fakhreddine S, Prommer H, Scanlon BR, Ying SC, Nicot J-P (2021) Mobilization of arsenic and other naturally occurring contaminants during managed aquifer recharge: a critical review. Environ Sci Technol 55(4):2208–2223. https://doi.org/10.1021/acs.est.0c07492

Fathi S, Hagen JSA, Haidari AH (2020) Synthesizing existing frameworks to identify the potential for managed aquifer recharge in a karstic and semi-arid region using GIS multi-criteria decision analysis. Groundw Sustain Dev:100390. https://doi.org/10.1016/j.gsd.2020.100390

Faunt CC, Sneed M, Traum J, Brandt JT (2016) Water availability and land subsidence in the Central Valley, California, USA. Hydrogeol J 24(3):675–684

Ganot Y, Holtzman R, Weisbrod N, Russak A, Katz Y, Kurtzman D (2018) Geochemical processes during managed aquifer recharge with desalinated seawater. Water Resour Res 54(2):978–994. https://doi.org/10.1002/2017WR021798

Ghasemizade M, Asante KO, Petersen C, Kocis T, Dahlke HE, Harter T (2019) An integrated approach toward sustainability via groundwater banking in the southern Central Valley, California. Water Resour Res 55(4):2742–2759

Government of India. Ministry of Water Resources (2012). National Water Policy. Adopted by National Water Resources Council 28 Dec 2012. http://wrmin.nic.in/writereaddata/NationalWaterPolicy/NWP2012Eng6495132651.pdf

Grinshpan M, Furman A, Dahlke HE, Raveh E, Weisbrod N (2021) From managed aquifer recharge to soil aquifer treatment on agricultural soils: concepts and challenges. Agric Water Manag 255:106991. https://doi.org/10.1016/j.agwat.2021.10699

Guppy L, Uyttendaele P, Villholth KG, Smakhtin VU (2018) Groundwater and sustainable development goals: analysis of interlinkages. UNU-INWEH Report Series

Ingrao C, Strippoli R, Lagioia G, Huisingh D (2023) Water scarcity in agriculture: an overview of causes, impacts and approaches for reducing the risks. Heliyon 9(8)

Jakeman AJ, Barreteau O, Hunt RJ, Rinaudo JD, Ross A, Arshad M, Hamilton S (2016) Integrated groundwater management: an overview of concepts and challenges. In: Integrated groundwater management: concepts, approaches and challenges. Springer, pp 3–20

Jasechko S, Perrone D (2021) Global groundwater wells at risk of running dry. Science 372(6540):418–421

Kazner C, Wintgens T, Dillon P (eds) (2012) Water reclamation technologies for safe managed aquifer recharge. IWA Publishing

Kim RH, Lee S, Lee JH, Kim YM (2007) Design of rainwater management system for eco-housing complex. In: Rainwater and urban design 2007. Engineers Australia, Barton, ACT, pp 634–639

Korkut M, Hartog N, Yavuz V (2025) An overview of historical development, current situation, and future prospects of managed aquifer recharge in Türkiye. Water 17(3):439

Kumar M, Panday DP, Goswami R, Neog N, Bahukhandi KD, Zhang J, Bhattacharya P (2025) Dynamics of arsenic co-contamination in the groundwater of the eastern Himalayas: the biogeochemical drivers and GRACE-based evidence. ACS ES & T Water. https://doi.org/10.1021/acsestwater.4c00106. (IF: 4.3; Q1)

la Cecilia D, Philipp M, Kaegi R, Schirmer M, Moeck C (2024) Microplastics attenuation from surface water to drinking water: impact of treatment and managed aquifer recharge—and identification uncertainties. Sci Total Environ 908:168378. https://doi.org/10.1016/j.scitotenv.2023.168378

Li W, Li Y, Long Y (2013) A review of research on artificial groundwater recharge wells. J Hohai Univ (Nat Sci) 41:410–416

Maliva R, Missimer T (2012) Managed aquifer recharge. In: Arid lands water evaluation and management. Springer Berlin Heidelberg, Berlin/Heidelberg, pp 559–630

Masoner JR, Kolpin DW, Cozzarelli IM, Barber LB, Burden DS, Foreman WT et al (2019) Urban stormwater: an overlooked pathway of extensive mixed contaminants to surface and groundwaters in the United States. Environ Sci Technol 53(17):10070–10081

Moeck C, Grech-Cumbo N, Podgorski J, Bretzler A, Gurdak JJ, Berg M, Schirmer M (2020) A global-scale dataset of direct natural groundwater recharge rates: a review of variables, processes and relationships. Sci Total Environ 717:137042. https://doi.org/10.1016/j.scitotenv.2020.137042

Mohammad-Hosseinpour A, Molina JL (2022) Improving the sustainability of urban water management through innovative groundwater recharge system (GRS). Sustainability 14(10):5990

Mukherjee A, Bhanja SN, Wada Y (2018) Groundwater depletion causing reduction of baseflow triggering Ganges river summer drying. Sci Rep 8(1):1–9

O'Geen A, Saal M, Dahlke H, Doll D, Elkins R, Fulton A, Fogg G, Harter T, Hopmans J, Ingels C (2015) Soil suitability index identifies potential areas for groundwater banking on agricultural lands. Calif Agric 69(2):75–84

Pachauri RK, Allen MR, Barros VR, Broome J, Cramer W, Christ R et al (2014) Climate change 2014: synthesis report. Contribution of Working Groups I, II and III to the fifth assessment report of the Intergovernmental Panel on Climate Change. IPCC, p 151

Page D, Bekele E, Vanderzalm J, Sidhu J (2018) Managed aquifer recharge (MAR) in sustainable urban water management. Water 10(3):239

Panday DP, Kumari A, Kumar M (2025) Alkalinity-salinity-sustainability: decadal groundwater trends and its impact on agricultural water quality in the Indian Peninsula. Sci Total Environ 978:179459

Pyne D (2005) Aquifer storage recovery: a guide to groundwater recharge through wells. ASR Systems, Gainesville

Ross CW, Prihodko L, Anchang J, Kumar S, Ji W, Hanan NP (2018) HYSOGs250m, global gridded hydrologic soil groups for curve-number-based runoff modeling. Sci Data 5:180091. https://doi.org/10.1038/sdata.2018.91

Sasidharan S, Bradford SA (2025) Assessing drywell designs for managed aquifer recharge via canals and repurposed wells. Sci Rep 15(1):1829

Scanlon BR, Fakhreddine S, Rateb A, de Graaf I, Famiglietti J, Gleeson T, Grafton RQ, Jobbagy E, Kebede S, Kolusu SR, Konikow LF, Long D, Mekonnen M, Schmied HM, Mukherjee A, MacDonald A, Reedy RC, Shamsudduha M, Simmons CT, Zheng C (2023) Global water resources and the role of groundwater in a resilient water future. Nat Rev Earth Environ. https://doi.org/10.1038/s43017-022-00378-6

Sprenger C, Hartog N, Hernández M, Vilanova E, Grützmacher G, Scheibler F, Hannappel S (2017) Inventory of managed aquifer recharge sites in Europe: historical development, current situation and perspectives. Hydrogeol J 25(6):1909–1922

Standen K, Costa LR, Monteiro JP (2020) In-channel managed aquifer recharge: a review of current development worldwide and future potential in Europe. Water 12(11):3099

Stefan C, Ansems N (2017) Web-based global inventory of managed aquifer recharge applications. Sustain Water Resour Manag 4(2):153–162. https://doi.org/10.1007/s40899-017-0212-6

Steinel A, Schelkes K, Subah A, Himmelsbach T (2016) Spatial multi-criteria analysis for selecting potential sites for aquifer recharge via harvesting and infiltration of surface runoff in North Jordan. Hydrogeol J 24(7):1753–1774

Sufyan M, Martelli G, Teatini P, Cherubini C, Goi D (2024) Managed aquifer recharge for sustainable groundwater management: new developments, challenges, and future prospects. Water 16(22):1–28

United Nations (2015) Transforming our world: the 2030 agenda for sustainable development. United Nations, New York. https://sustainabledevelopment.un.org/post2015/transformingourworld

Villholth KG, Sood A, Liyanage N, Zhu T, Wada Y (2016) Global food production—share from sustainable and unsustainable groundwater use. In: 43rd IAH Congress, le Corum, Montpellier, France, 25–29 September 2016

WHO (2011) Guidelines for drinking water quality, 4th edn. World Health Organisation, Geneva

Zhang H, Xu Y, Kanyerere T (2020) A review of the managed aquifer recharge: historical development, current situation, and perspectives. Phys Chem Earth Parts A/B/C 118:102887

Chapter 4
Trends in Geogenic Contaminants and Groundwater Exploration

Abstract Geogenic contamination of groundwater has become one of the most pervasive global water-quality challenges, affecting nearly one billion people across diverse hydrogeologic and climatic settings. This review synthesizes three decades of empirical and modeling studies on the occurrence, mobility, and management of major geogenic contaminants—arsenic (As), fluoride (F^-), iron (Fe), manganese (Mn), and uranium (U)—which frequently exceed WHO guideline limits of 10 μg L^{-1} for As, 1.5 mg L^{-1} for F^-, and 30 μg L^{-1} for U. Spatial datasets show that As contamination (>10 μg L^{-1}) affects over 200 million people in deltaic Asia. In contrast, F^- concentrations up to 9 mg L^{-1} are reported in India and East Africa, and U above 0.03 mg L^{-1} occurs in more than 30% of tested wells in Punjab, India. These contaminants originate primarily from rock–water interactions—reductive dissolution, alkaline weathering, and carbonate complexation—but are now strongly modulated by groundwater abstraction, irrigation return flow, and managed aquifer recharge. The synthesis highlights how machine-learning models (e.g., Random Forest, SVM) coupled with isotopic and reactive-transport simulations are reshaping predictive groundwater assessment. Such hybrid frameworks have achieved high classification accuracy in identifying high-risk zones in China, India, and Africa, transforming traditional exploration into proactive hazard mapping. Integrating global geochemical, geospatial, and land-use datasets further reveals that contaminant hotspots coincide with irrigated croplands and peri-urban growth zones, underscoring the anthropogenic amplification of geogenic risks. The review concludes that sustainable management must move from reactive treatment toward preventive aquifer stewardship, embedding predictive analytics, geochemical compatibility design, and participatory governance within groundwater policy.

Keywords Geogenic contamination · Groundwater exploration · Hydrogeochemical modeling · Arsenic and fluoride · Machine learning

M. Kumar, D. P. Panday, *Enabling Water Security and Sustainability: Global Challenges and Solutions*, SpringerBriefs in Water Science and Technology, https://doi.org/10.1007/978-3-032-21521-5_4

4.1 Introduction

Groundwater represents the largest accessible freshwater reservoir on Earth, supplying drinking water to nearly half of the global population and supporting agricultural, industrial, and ecological systems (Smedley and Kinniburgh 2002). However, its chemical quality is increasingly threatened by both anthropogenic and natural factors. Among these, geogenic contaminants—elements mobilized from aquifer rocks and sediments through natural geochemical processes—pose a persistent and often underrecognized challenge to sustainable groundwater management. Unlike anthropogenic pollutants, geogenic contaminants arise from mineral dissolution, weathering, ion exchange, and redox reactions within aquifers. Arsenic (As), fluoride (F^-), iron (Fe), manganese (Mn), and uranium (U) are the most widespread geogenic contaminants and frequently exceed drinking-water standards across diverse hydrogeological settings (Nickson et al. 1998; Fendorf et al. 2010; Podgorski and Berg 2020).

Arsenic contamination is among the most severe geogenic groundwater crises worldwide. Since its identification in Bangladesh in the late 1990s (Nickson et al. 1998), arsenic has been documented across South and Southeast Asia, China, and Latin America (Rodríguez-Lado et al. 2013; Singh et al. 2018; Mahlknecht et al. 2022). Its mobilization is primarily controlled by redox-driven dissolution of iron oxyhydroxides and competitive ion-exchange processes under reducing aquifer conditions (Smedley and Kinniburgh 2002; Bhattacharya et al. 2019). Fluoride contamination, in contrast, is prevalent in arid and semi-arid crystalline terrains of East Africa, India, and northern Mexico, where alkaline conditions, long residence times, and high evapotranspiration promote mineral dissolution (Amini et al. 2008; Armienta and Segovia 2008; Rango et al. 2014). Elevated Fe and Mn concentrations similarly reflect reducing conditions that destabilize metal oxyhydroxides (Knappett et al. 2017; Zhang et al. 2022). More recently, uranium has emerged as a significant geogenic contaminant in alluvial and granitic aquifers of northwestern India and the western United States, highlighting the role of carbonate equilibria and redox complexity in contaminant mobility (Shukla et al. 2021; Smedley et al. 2022).

Although these processes are natural, their expression in groundwater systems is increasingly shaped by human activities. Intensive groundwater abstraction alters hydraulic gradients and redox zonation, enhancing contaminant mobilization and transport (Mukherjee et al. 2008; Fendorf et al. 2010). Land-use change, irrigation return flows, and managed aquifer recharge (MAR) further modify water–rock interactions and geochemical equilibria, sometimes triggering unintended contaminant release (Yang et al. 2015; Fakhreddine et al. 2024). These coupled natural–anthropogenic controls necessitate integrated approaches that bridge hydrogeology, geochemistry, and environmental engineering.

Advances in groundwater exploration and assessment have paralleled growing awareness of geogenic contamination. Traditional hydrogeological methods are now complemented by geophysical, isotopic, and geospatial tools, while machine learning (ML) and remote sensing enable high-resolution prediction of contamination risk (Rodríguez-Lado et al. 2013; Podgorski and Berg 2020). Statistical and ML-based models, including Random Forests and Support Vector Machines, are widely applied for arsenic and fluoride risk mapping, while geochemical and reactive transport models (e.g., PHREEQC, MT3DMS) provide mechanistic insight into mobilization processes (Souza et al. 2017; Hasan et al. 2018). Together, these approaches represent a shift toward predictive groundwater quality assessment.

Mitigation strategies increasingly integrate hydrochemical understanding with engineering and governance solutions. While treatment technologies such as granular ferric hydroxide, activated alumina, and biological filtration are effective at local scales (Thirunavukkarasu et al. 2003; Zhang et al. 2024), long-term sustainability depends on source substitution, aquifer management, and adaptive policy frameworks (Hussam and Munir 2007; Mukherjee and Bhattacharya 2022). Despite extensive research, challenges remain, including substantial spatial variability, incomplete understanding of coupled biogeochemical processes, and limited transferability of predictive models across geological settings.

The objective of this review is to synthesize global patterns, mechanisms, and trends of geogenic contaminants in groundwater and to evaluate the evolution of exploration and predictive techniques underpinning their assessment. Specifically, it examines (i) governing geochemical and hydrogeological processes, (ii) global spatial and temporal trends of major contaminants, (iii) advances in traditional and emerging exploration methods, and (iv) sustainable mitigation frameworks that balance groundwater development with long-term quality protection.

4.2 Geochemical and Hydrogeological Controls on Geogenic Contamination

Groundwater chemistry is governed by interactions among lithology, mineralogy, hydrodynamics, and redox conditions, which together control the occurrence and mobility of geogenic contaminants. Arsenic, fluoride, iron, manganese, and uranium originate from rock–water interactions that release trace elements through mineral dissolution, weathering, ion exchange, and redox reactions. Unlike surface-derived pollution, geogenic contamination is typically time-dependent and evolves over decades as aquifer conditions change (Smedley and Kinniburgh 2002; Fendorf et al. 2010). Understanding these processes is essential for predicting contaminant distribution and informing sustainable groundwater exploration.

4.2.1 Geochemical Processes Governing Contaminant Mobilization

Arsenic mobilization is the most extensively studied geogenic process due to widespread public health impacts in South and Southeast Asia. In alluvial aquifers of the Bengal Basin, arsenic is released via microbially mediated reductive dissolution of

Fe(III) oxyhydroxides that host adsorbed arsenic (Nickson et al. 1998; Bhattacharya et al. 2019). Organic carbon from sediments or recharge fuels microbial respiration, driving redox progression from oxygen and nitrate reduction to iron reduction, thereby liberating arsenic and elevating dissolved iron concentrations. Groundwater abstraction alters flow paths and redox boundaries, redistributing arsenic within aquifers and linking its occurrence to anthropogenic water use (Mukherjee et al. 2008; Fendorf et al. 2010; Panday et al. 2024).

Fluoride contamination follows a contrasting alkaline dissolution pathway. In arid and semi-arid regions such as northern Mexico, Rajasthan, and the East African Rift, high evapotranspiration, long groundwater residence times, and low calcium concentrations favor dissolution of fluorite, apatite, and biotite (Amini et al. 2008; Armienta and Segovia 2008; Rango et al. 2014). Elevated bicarbonate and alkaline pH enhance fluoride solubility through cation exchange and complexation, while limited recharge restricts dilution, frequently resulting in concentrations exceeding the WHO guideline of 1.5 mg L^{-1}. Climatic variability further modulates these processes, with drought enhancing evaporative concentration and episodic recharge providing transient dilution.

Redox dynamics likewise control iron and manganese contamination. Under reducing conditions driven by microbial organic matter degradation, Fe(III) and Mn(IV) oxyhydroxides dissolve, releasing Fe^{2+} and Mn^{2+} into groundwater (Knappett et al. 2017; Zhang et al. 2022). Because these phases act as key sorbents for arsenic, their reduction often mobilizes multiple contaminants simultaneously. Seasonal flooding, recharge, and anthropogenic disturbances cause redox oscillations, producing substantial spatial and temporal variability in Fe–Mn concentrations. Although not acutely toxic at moderate levels, chronic manganese exposure has been linked to neurodevelopmental effects.

Uranium mobility illustrates the sensitivity of geogenic contamination to oxidation state and carbonate chemistry, in oxidizing, bicarbonate-rich aquifers, uranium forms soluble uranyl–carbonate complexes, enhancing transport (Shukla et al. 2021; Smedley et al. 2022; Panday and Kumar 2024). Under reducing conditions, U(VI) may be immobilized as U(IV); however, nitrate infiltration or MAR using oxygenated water can reverse reduction and remobilize uranium (Yang et al. 2015). Evidence from Punjab, India, and the western United States shows that such geochemical oscillations can elevate uranium above 30 μg L^{-1}, posing chronic radiological risks (CGWB 2025).

4.2.2 Hydrogeological Architecture, Microbial Processes, and Climate Influences

Hydrogeological structure strongly governs the spatial manifestation of geogenic processes. Fine-grained, organic-rich Holocene sediments favor reducing conditions conducive to arsenic and iron mobilization, whereas older Pleistocene sands tend to remain more oxidized (Mukherjee et al. 2008; Fendorf et al. 2010; Panday et al. 2025). Crystalline and volcanic aquifers, despite lower transmissivity, commonly host fluoride and uranium due to mineral composition and extended water–rock interaction times. Mixing between chemically distinct aquifers can trigger secondary reactions, including redox buffering and cycles of mineral precipitation and dissolution.

Hydrologic fluxes further regulate contaminant persistence and transport. High recharge may dilute solutes but can also introduce electron acceptors that oxidize reduced sediments, releasing sorbed contaminants (Hasan et al. 2018). Conversely, over-extraction depresses water tables and promotes downward migration of contaminated shallow groundwater into deeper aquifers, particularly where confining layers are breached by pumping or poorly constructed wells (Fendorf et al. 2010). Such vertical migration has been documented in the Bengal Basin and the Comarca Lagunera, Mexico (Mahlknecht et al. 2022; Ahirvar et al. 2025). Microbial processes play a central role in mediating these transformations. Microorganisms catalyze redox reactions involving Fe, Mn, As, and S, linking carbon cycling directly to contaminant mobilization (Bhattacharya et al. 2019).

Global studies demonstrate how geogenic groundwater contamination is governed by coupled geochemical, hydrogeological, and anthropogenic controls, and how exploration, modeling, and mitigation strategies have evolved accordingly. Arsenic (As) contamination has been most extensively documented in South and Southeast Asia, particularly in the alluvial aquifers of the Bengal Basin, where reductive dissolution under anoxic conditions releases arsenic from iron oxyhydroxides (Nickson et al. 1998; Smedley and Kinniburgh 2002; Fendorf et al. 2010; Bhattacharya et al. 2019). Groundwater abstraction alters flow paths and redox fronts, redistributing arsenic spatially and temporally (Mukherjee et al. 2008; Fendorf et al. 2010; Hasan et al. 2018). Similar redox-linked mobilization of Fe and Mn has been reported across South and Southeast Asia and in glacial aquifers of the United States, where Mn remains under-monitored despite chronic health concerns (Knappett et al. 2017; Zhang et al. 2022).

Global arsenic risk mapping has advanced from conceptual hydrogeological interpretation to quantitative prediction using machine learning. Random Forest–based models have identified high-risk arsenic belts in South Asia, China, and previously undocumented regions, estimating tens of millions of people at risk (Rodríguez-Lado et al. 2013; Podgorski and Berg 2020; Zhang et al. 2014). These data products now underpin mitigation-targeting and groundwater-testing programs worldwide (Eawag Open Data; Podgorski and Berg 2020). Fluoride (F^-) contamination shows a contrasting but equally widespread pattern, concentrated in arid and semi-arid volcanic and sedimentary terrains of East Africa, India, China, and northern Mexico, where alkaline conditions, evapoconcentration, and long residence times promote mineral dissolution (Amini et al. 2008; Armienta and Segovia 2008; Rango et al. 2014; Ijumulana et al. 2021). Regional studies from Ethiopia, Tanzania, Odisha, and Comarca Lagunera document strong links between groundwater fluoride and fluorosis prevalence, necessitating defluoridation or source substitution (Rango et al. 2014; Mahlknecht et al. 2022; SPCB Odisha/TOI 2025).

Mitigation strategies range from source management and well-depth optimization to household and community treatment. In Bangladesh and India, shallow–deep well switching, subsurface arsenic removal (SAR), and community arsenic removal units (ARUs) have proven effective when supported by appropriate operation and maintenance (Van Geen et al. 2008; Rahman et al. 2015; Mukherjee and Bhattacharya 2022). Point-of-use technologies—including SONO composite iron filters, granular ferric hydroxide, activated alumina, bone char, and membrane processes—have demonstrated arsenic and fluoride removal to below guideline values, though scalability and long-term sustainability remain challenges (Thirunavukkarasu et al. 2003; Hussam and Munir 2007; Ayoob and Gupta 2006, 2008; Shen et al. 2020; Choubisa 2023). For Fe and Mn, aeration, biological rapid sand filtration, and Mn-modified media have achieved high removal efficiencies in groundwater treatment plants (Niu et al. 2022; Zhang et al. 2024).

Uranium (U) has emerged as a critical geogenic contaminant in alluvial and granitic aquifers of northwestern India and the western United States, where carbonate complexation under oxidizing conditions enhances mobility (Shukla et al. 2021; Smedley et al. 2022). Monitoring reports from Punjab indicate that over 30% of samples exceed the 30 $\mu g\ L^{-1}$ guideline, often co-occurring with fluoride and arsenic, necessitating advanced treatment or alternative sources. Reactive transport and speciation modeling (PHREEQC, GWB, MODFLOW–MT3DMS) increasingly inform both risk prediction and mitigation design, including MAR schemes where recharge chemistry can inadvertently mobilize arsenic or uranium if not carefully managed (Souza et al. 2017; Yang et al. 2015; Fakhreddine et al. 2024).

Metagenomic studies increasingly identify functional genes associated with arsenate reduction and sulfate respiration, emphasizing the need for coupled hydro-biogeochemical models. Climatic setting further modulates geogenic pathways: arid climates enhance fluoride enrichment through evapoconcentration, while humid monsoonal systems promote reducing conditions favorable for arsenic release (Amini et al. 2008; Fendorf et al. 2010). Climate change is likely to intensify both processes through prolonged droughts and increased extreme rainfall.

4.3 Global Distribution and Emerging Trends of Geogenic Contaminants

The global distribution of geogenic contaminants reflects the interaction of lithological inheritance, hydrological evolution, climate, and human intervention. Although governed by natural geochemical processes, contamination is spatially heterogeneous, ranging from arsenic-rich alluvial basins in Asia to fluoride-dominated crystalline aquifers in Africa and Latin America. Over the past three decades, large-scale mapping, statistical modeling, and machine-learning (ML) approaches have transformed understanding of these patterns, enabling quantitative global risk assessments (Smedley and Kinniburgh 2002; Podgorski and Berg 2020).

4.3.1 *Arsenic: The Paradigm of Geogenic Contamination*

Arsenic remains the most intensively studied geogenic contaminant and a significant global public health concern. Its discovery in the Bengal Basin in the late 1990s (Nickson et al. 1998) prompted extensive investigations that revealed widespread contamination across South and Southeast Asia, China, and Latin America (Fendorf et al. 2010; Rodríguez-Lado et al. 2013; Mahlknecht et al. 2022). In the Bengal Basin, arsenic concentrations frequently exceed 50 μg L^{-1} due to microbially mediated reductive dissolution of Fe(III) oxyhydroxides in organic-rich Holocene sediments (Bhattacharya et al. 2019). Spatial heterogeneity is controlled by fluvio-deltaic stratigraphy, sediment age, and organic carbon availability, while intensive irrigation pumping has altered flow paths and redox boundaries, promoting vertical arsenic migration (Mukherjee et al. 2008; Fendorf et al. 2010; Agarwal et al. 2024). Beyond South Asia, arsenic contamination affects northern China, where statistical risk models estimate nearly 20 million people exposed above WHO guidelines (Rodríguez-Lado et al. 2013). In arid alluvial basins such as Comarca Lagunera, Mexico, arsenic co-occurs with fluoride under evaporative and irrigation-influenced conditions (Armienta and Segovia 2008; Mahlknecht et al. 2022). Additional enrichment occurs in volcanic and geothermal terrains of the southwestern United States and Latin America (Smedley and Kinniburgh 2002).

4.3.2 *Fluoride: The Silent Epidemic of Arid Landscapes*

Fluoride contamination is geographically more widespread than arsenic contamination, particularly in semi-arid crystalline and volcanic aquifers in India, East Africa, China, and northern Mexico. Its mobilization results from prolonged interaction between alkaline groundwater and fluoride-bearing minerals such as fluorite, apatite, and biotite (Amini et al. 2008; Rango et al. 2014). Global geospatial analyses identify fluoride hotspots associated with low calcium, high pH, and arid climates, demonstrating strong predictive capability (Amini et al. 2008). Chronic exposure above 1.5 mg L^{-1} causes dental and skeletal fluorosis, which is now endemic in several Indian states and the African rift systems (Rango et al. 2014; Choubisa 2023). In Mexico, fluoride enrichment correlates with evapotranspiration intensity and groundwater residence time, underscoring climatic control (Armienta and Segovia 2008). Recent studies highlight climate feedbacks: warming and aridification intensify evaporative concentration, whereas episodic recharge provides only transient dilution without reversing long-term accumulation (Ijumulana et al. 2021).

4.3.3 Iron and Manganese: Indicators of Reductive Aquifer Dynamics

Iron (Fe) and manganese (Mn) commonly co-occur with arsenic and serve as indicators of reducing conditions in aquifers. Under anoxic environments, reductive dissolution of Fe–Mn oxyhydroxides releases Fe^{2+} and Mn^{2+}, often coinciding with arsenic mobilization (Knappett et al. 2017; Zhang et al. 2022). Elevated Mn, once considered primarily an aesthetic issue, is now linked to chronic neurodevelopmental effects. Global syntheses reveal widespread Mn enrichment in shallow anoxic aquifers across Asia, Africa, and North America, frequently overlooked in monitoring programs (Zhang et al. 2022). These redox-linked associations underpin statistical and reactive transport models that use Fe and Mn as proxies for arsenic risk prediction (Podgorski and Berg 2020).

4.3.4 Uranium and Radionuclides: The Emerging Frontier

Uranium is an emerging geogenic contaminant that has been increasingly documented in alluvial and granitic aquifers of northwestern India, the western United States, and parts of Europe (Shukla et al. 2021; Smedley et al. 2022). In oxidizing, bicarbonate-rich systems, uranium forms mobile uranyl–carbonate complexes, whereas in reducing conditions it immobilizes as U(IV) phases (Smedley et al. 2022). Anthropogenic drivers—including irrigation, nitrate infiltration, and groundwater abstraction—can shift redox equilibria and enhance uranium mobility. A state-wide survey in Punjab, India, showed that 32% of samples exceeded 0.03 mg L^{-1}, often co-occurring with arsenic and fluoride, indicating multi-contaminant risk (CGWB 2025).

4.3.5 Climate, Land Use, and Dynamic Controls

Although geogenic contamination is rooted in geology, its distribution is strongly modified by climate variability and land use. Droughts and prolonged pumping lower water tables and oxidize reduced sediments, remobilizing uranium and manganese, whereas intense rainfall and flooding introduce organic-rich recharge that stimulates arsenic release (Fendorf et al. 2010; Yang et al. 2015; Fakhreddine et al. 2024). Land-use change, irrigation, and managed aquifer recharge (MAR) alter recharge chemistry and equilibrium states; MAR using oxygenated or nitrate-rich water can inadvertently mobilize arsenic and uranium unless geochemical compatibility is ensured (Yang et al. 2015; Fakhreddine et al. 2024). These temporal oscillations render geogenic contamination a dynamic rather than static hazard. Figure 4.1 synthesizes global distribution patterns of fluoride (F^-), arsenic (As), and uranium (U) alongside land-use and land-cover (LULC) data. Fluoride hotspots (Fig. 4.1c) align with semi-arid and volcanic provinces of Asia, Africa, and Latin America, reflecting alkaline conditions, low calcium, and evapoconcentration (Podgorski and Berg 2022). Arsenic distribution (Fig. 4.1b) is concentrated in Holocene alluvial and deltaic plains, such as the Ganges–Brahmaputra–Meghna Basin, the Red River Delta, and the North China Plain, where organic-rich sediments favor reductive dissolution (Mukherjee et al. 2024). Uranium enrichment (Fig. 4.1b) occurs in oxidizing, carbonate-rich aquifers of northwestern India, the western United States, Australia, and northern China, often coinciding with intensive groundwater abstraction (Riedel and Kübeck 2018).

The LULC overlay (Fig. 4.1d; Jing et al. 2024) reveals substantial spatial overlap between contaminated aquifers and cropland, rangeland, and peri-urban zones, highlighting the amplification of geogenic processes by agricultural intensification, irrigation return flow, and urbanization. In contrast, forested and wetland-dominated regions exhibit lower contaminant prevalence, underscoring the buffering role of stable recharge and natural land cover. Together, Fig. 4.1 provides a spatial

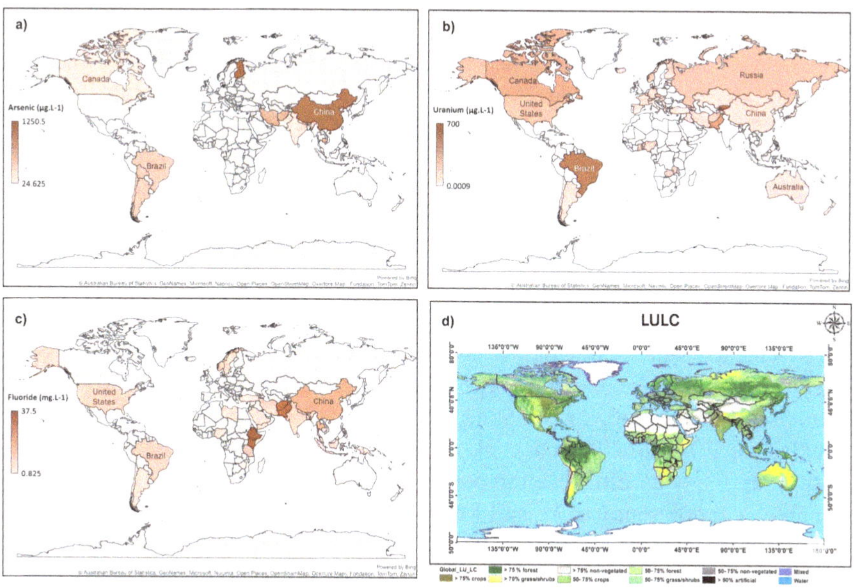

Fig. 4.1 Global distribution of major geogenic contaminants and land-use patterns. Spatial maps showing the worldwide occurrence of (**a**) arsenic (As), (**b**) uranium (U), and (**c**) fluoride (F^-) in groundwater, together with (**d**) global land-use and land-cover (LULC) patterns. The maps illustrate how contaminant hotspots align with specific hydrogeologic settings—reductive alluvial basins for As, oxidizing granitic and volcanic terrains for U, and arid alkaline regions for F^-—and overlap substantially with cropland and peri-urban expansion zones. Darker shades represent higher concentrations or greater land-use intensity, highlighting areas of compounded geogenic and anthropogenic stress on groundwater quality. **Data sources:** Compiled from Mukherjee et al. (2024) for As, Podgorski and Berg (2022) for F^-, Riedel and Kübeck (2018) for U, and Jing et al. (2024) for Land use land cover (LULC)

synthesis linking lithology, climate, and human activity, reinforcing the need for integrated hydrogeochemical, geospatial, and ML-based approaches to anticipate and manage geogenic groundwater contamination.

4.3.6 Spatial Prediction and the Role of Machine Learning

Advances in machine learning have revolutionized the spatial prediction of geogenic contaminants. Random Forest and boosted regression tree models now integrate geological, geochemical, climatic, and topographic predictors to produce global probability maps of arsenic and fluoride occurrence (Rodríguez-Lado et al. 2013; Podgorski and Berg 2020). These models, validated using extensive well datasets, identify high-risk regions that previously lacked field data. ML approaches have also been applied at regional scales to delineate fluoride-prone zones in the Ethiopian Rift and uranium hotspots in Indian alluvial aquifers (Shukla et al. 2021; Ijumulana et al. 2021). Beyond mapping, data-driven frameworks provide insight into process hierarchies—highlighting, for example, the dominant influence of redox proxies and aquifer lithology over climate variables. However, while predictive accuracy is improving, the interpretability of ML models remains a challenge for hydrogeochemists, underscoring the need for hybrid frameworks that combine statistical learning with mechanistic modeling.

4.3.7 Co-contamination and Multi-Element Risk Zones

A striking feature of geogenic contamination is the frequent co-occurrence of multiple elements, which compounds health risks and analytical challenges. Arsenic often coexists with Fe and Mn under reducing conditions, while fluoride and uranium coincide in oxidizing, alkaline environments (Armienta and Segovia 2008; Shukla et al. 2021; Mahlknecht et al. 2022). This spatial overlap complicates remediation, as methods effective for one contaminant may exacerbate another. For example, aeration removes Fe and Mn but can oxidize arsenic-bearing sediments, increasing dissolved As. Similarly, reverse osmosis removes fluoride but produces concentrated brine that must be disposed of safely. Recognizing the hydrochemical typologies of multi-contaminant zones is therefore essential for designing appropriate treatment and management strategies. Figure 4.2 highlights the evolution of geogenic contaminants and the global hotspots.

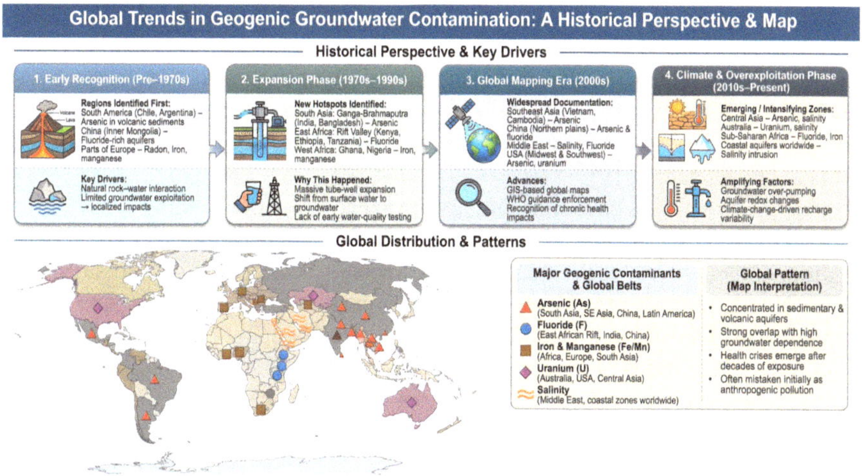

Fig. 4.2 Technological and conceptual evolution of Managed Aquifer Recharge (MAR) from ancient empirical practices to modern smart-integrated systems. The timeline illustrates a paradigm shift from simple community-based water harvesting (Ancient–1950s) to sophisticated hydrogeochemical characterization using environmental isotopes (2010–2018) and the current integration of machine learning and real-time monitoring. The transition highlights the increasing role of MAR in the urban water circular economy and climate-resilient planning

4.4 Advances in Groundwater Exploration and Predictive Modelling

Groundwater exploration has progressed from descriptive hydrogeological surveys to an integrated, multi-disciplinary framework combining geophysics, hydrogeochemistry, isotopic tracing, and machine learning (ML). This evolution reflects the recognition that groundwater quantity and quality are inseparable, requiring exploration strategies that simultaneously assess aquifer potential and vulnerability to geogenic contamination (Fendorf et al. 2010; Podgorski and Berg 2020).

Conventional exploration methods—hydrogeological mapping, borehole logging, and well inventories—remain essential for characterizing aquifer geometry, transmissivity, and recharge, but they offer limited predictive power for spatially heterogeneous contaminants such as arsenic and fluoride. Consequently, geophysical techniques including electrical resistivity tomography (ERT), electromagnetic (EM) induction, and ground-penetrating radar are increasingly applied to delineate subsurface stratigraphy and identify clay-rich, saline, or reducing zones that influence contaminant transport (Mukherjee et al. 2008). In deltaic aquifers, ERT has proven particularly effective in distinguishing oxidized from reduced sediments and inferring arsenic susceptibility.

Hydrogeochemical analysis remains central to exploration, with classical tools such as Piper, Gibbs, and Durov diagrams used to identify groundwater facies and dominant geochemical processes (Smedley and Kinniburgh 2002). Multivariate statistical techniques—principal component analysis, factor analysis, and clustering—further resolve complex ion-chemistry patterns linked to weathering, ion exchange, or mixing (Mahlknecht et al. 2022). When integrated with isotopic tracers ($\delta^{2}H$, $\delta^{18}O$, $\delta^{13}C$, $\delta^{34}S$), these methods provide temporal and process-based insight into recharge dynamics and redox-driven contaminant mobilization (Bhattacharya et al. 2019).

Isotopic techniques have substantially advanced the understanding of groundwater origin, flow paths, and contaminant sources. Stable isotopes of oxygen and hydrogen distinguish local versus regional recharge, while carbon and sulfur isotopes elucidate redox reactions involving organic matter oxidation and sulfate reduction (Mukherjee and Bhattacharya 2022). Radiogenic uranium isotope ratios ($^{238}U/^{235}U$, $^{234}U/^{238}U$) are increasingly applied to assess uranium mobility under oxidizing conditions in carbonate-rich aquifers (Shukla et al. 2021). Coupled hydrochemical–isotopic approaches thus provide critical constraints for predictive modeling and remediation design.

4.4.1 Process-Based and Machine Learning Models

Advances in geochemical and reactive transport modeling have strengthened the mechanistic understanding of contaminant behavior. Tools such as PHREEQC, Geochemist's Workbench, and MINTEQ simulate equilibrium and kinetic reactions under variable pH–Eh conditions, while reactive transport models (e.g., MT3DMS, HYDRUS) couple chemical reactions with groundwater flow to evaluate time-dependent contaminant migration (Souza et al. 2017; Hasan et al. 2018). These models quantify sensitivity to redox shifts, bicarbonate levels, and organic carbon availability, supporting scenario-based risk assessment.

In parallel, ML approaches—including Random Forest, Support Vector Machines, and Gradient Boosting—have emerged as powerful tools for large-scale groundwater quality prediction. By integrating hydrogeochemical, lithological, climatic, and topographic variables, ML models capture nonlinear interactions often missed by classical statistics (Rodríguez-Lado et al. 2013; Podgorski and Berg 2020). National-scale applications in China and regional studies in India and Africa demonstrate the scalability of ML–GIS frameworks for arsenic, fluoride, and uranium risk mapping (Rodríguez-Lado et al. 2013; Ijumulana et al. 2021; Shukla et al. 2021).

4.4.2 Integration, Interpretability, and Emerging Trends

To enhance decision relevance, recent efforts emphasize hybrid modeling frameworks that integrate ML predictions with process-based geochemical models. Techniques such as SHAP and permutation importance improve interpretability by ranking the influence of predictors and ensuring consistency with hydrogeochemical principles. The integration of remote sensing and GIS further extends exploration capacity, using satellite-derived indicators of land surface temperature, vegetation, soil moisture, and lineament density as proxies for recharge and lithological control (Amini et al. 2008). Emerging trends include real-time monitoring using in situ sensors for redox potential, conductivity, and temperature, combined with ML-driven analytics to provide early warning of contaminant mobilization linked to hydrological events (Podgorski and Berg 2020). These advances shift groundwater exploration from static mapping toward dynamic risk forecasting.

Despite significant progress, challenges remain, including data scarcity, uneven analytical quality, limited model transferability, and institutional barriers to implementation—particularly in regions most affected by geogenic contamination. Addressing these constraints requires open data platforms, standardized modeling protocols, and participatory frameworks that link predictive science with groundwater governance. Collectively, these advances mark a paradigm shift toward predictive, integrative groundwater exploration that anticipates contaminant evolution and informs targeted mitigation strategies.

The integration of multi-disciplinary datasets, ranging from field-based geological observations to advanced remote sensing and machine learning algorithms, is essential for developing accurate predictive models of complex groundwater systems. These integrated frameworks enable proactive risk assessment and forecasting, providing crucial decision support for sustainable aquifer management and effective contaminant mitigation strategies (Fig. 4.3).

4.4.3 Mitigation, Treatment Technologies, and Future Research Directions

Managing geogenic contaminants like As, F^- and U is a delicate balancing act between technical precision and socio-political willpower. While we understand the geochemical triggers—such as reductive dissolution or alkaline desorption—translating this into lasting safety for vulnerable communities remains uneven. Often, the first line of defense is source substitution, such as tapping deeper Pleistocene sands in the Bengal Basin or importing surface water in arid Mexico and India (Fendorf et al. 2010; Mukherjee et al. 2008; Armienta and Segovia 2008). However, these are not permanent fixes; heavy pumping risks drawing contaminants downward, making constant monitoring and isotopic tracing vital to ensure "safe" zones don't become tomorrow's hazards (Mukherjee and Bhattacharya 2022; Mahlknecht et al. 2022).

Where substitution fails, we rely on a spectrum of treatment technologies. From community-scale SONO filters to the Nalgonda process for F, these systems succeed only when they align with local economic realities (Hussam 2009; Ayoob and Gupta 2008; Choubisa 2023). Adsorption filters using granular ferric hydroxide

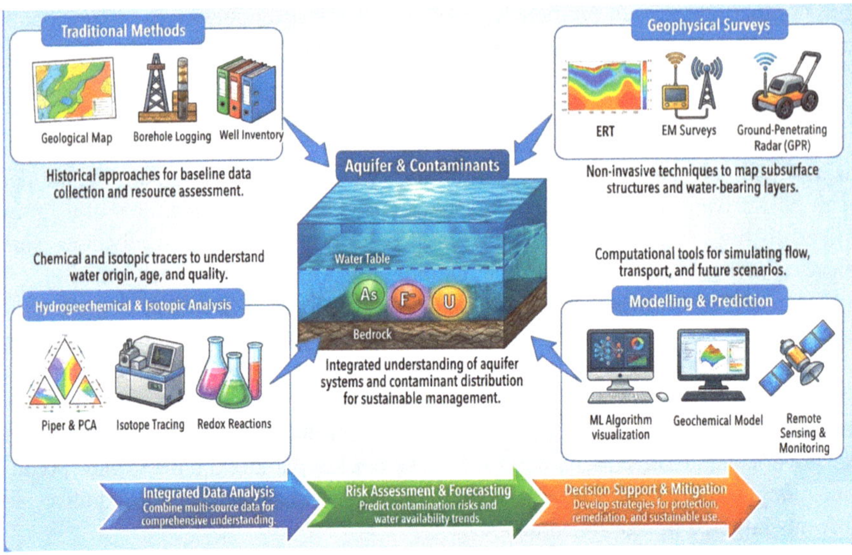

Fig. 4.3 Comprehensive overview of integrated approaches in modern groundwater exploration and predictive modelling. This diagram illustrates how diverse methodologies—including traditional field methods, non-invasive geophysical surveys, detailed hydrogeochemical analysis, and advanced computational modelling—are combined to characterize aquifer systems and the distribution of contaminants. This integrated data analysis supports robust risk assessment, forecasting, and informed decision-making for sustainable resource management and mitigation strategies

(GFH), activated alumina, or iron-oxide-coated sand show high efficiency, yet maintenance remains a barrier (Thirunavukkarasu et al. 2003; Hussam and Munir 2007; Zhang et al. 2024; Niu et al. 2022). We are now seeing a shift toward biological sand filters and bio-oxidation systems that use indigenous microbes to remove Fe and Mn without heavy chemical dosing (Zhang et al. 2024; Niu et al. 2022). At the aquifer scale, Managed Aquifer Recharge (MAR) and Subsurface Arsenic Removal (SAR) offer innovative ways to dilute pollutants in situ (Rahman et al. 2015). Yet, these carry geochemical risks; introducing oxygenated or carbon-rich water can inadvertently wake up dormant contaminants by shifting redox equilibria (Yang et al. 2015; Fakhreddine et al. 2024).

The future of groundwater safety lies in democratizing data. By integrating machine learning (ML), GIS-based risk mapping, and real-time field sensors, we can move from reactive crisis management to proactive stewardship (Rodríguez-Lado et al. 2013; Podgorski and Berg 2020). This digital frontier, combined with citizen science and integrated water observatories, empowers remote communities to participate in their own governance (Mukherjee and Bhattacharya 2022). Furthermore, our strategies must be climate-resilient; as droughts concentrate F and floods trigger As release, models must factor in hydroclimatic variables to safeguard the "human-magnified" reactive systems our aquifers have become (Amini et al. 2008; Ijumulana et al. 2021). Emerging techniques like isotopic fingerprinting and non-target analytical (NTA) methods, combined with portable X-ray fluorescence (pXRF), will further bridge the gap between monitoring and life-saving intervention (Bhattacharya et al. 2019).

4.5 Conclusion

The collective evidence demonstrates that geogenic contamination is a geological inheritance magnified by human hands. While As, F^-, and U mobilization follow established pathways, the magnitude of the current crisis is primarily determined by anthropogenic disturbances to natural redox and flow regimes. Excessive abstraction, irrigation return flow, and poorly designed recharge schemes have converted once-stable aquifers into dynamic, reactive systems in which even minor chemical perturbations trigger large-scale toxic releases. Ultimately, securing safe groundwater requires an anticipatory framework that unites scientific modeling with institutional capacity and community participation. We must move beyond local mitigation toward preventive aquifer stewardship, embedding geochemical insight into land-use policy and infrastructure design. By aligning predictive science with Sustainable Development Goal 6 (SDG 6), we can transition from reactive crisis response to a future of long-term environmental resilience and proactive governance.

References

Agarwal V, Kumar M, Panday DP, Zang J, Munoz-Arriola F (2024) Unlocking the potential of remote sensing for arsenic contamination detection and management: challenges and perspectives. Curr Opin Environ Sci Health:100578. https://doi.org/10.1016/j.coesh.2024.100578

Ahirvar BP, Das P, Kumar R, Panday DP, Das N, Srivastva PK, Kumar M (2025) Weathering in the subsurface environment of the Gondwana deposits intensified by the cumulative impacts of mining and agriculture practices. J Contam Hydrol:104630. https://doi.org/10.1016/j.jconhyd.2025.104630

Amini M, Mueller K, Abbaspour KC, Rosenberg T, Afyuni M, Møller KN, Sarr M, Johnson CA (2008) Statistical modeling of global geogenic fluoride contamination in groundwaters. Environ Sci Technol 42(10):3662–3668

Armienta MA, Segovia N (2008) Arsenic and fluoride in the groundwater of Mexico. Environ Geochem Health 30(4):345–353

Ayoob S, Gupta AK (2006) Fluoride in drinking water: a review on the status and stress effects. Crit Rev Environ Sci Technol 36(6):433–487

Ayoob S, Gupta AK (2008) Defluoridation techniques for drinking water and their applicability in developing countries. J Environ Manag 90(2):552–561

Bhattacharya P, Kumar A, Mukherjee A (2019) Conceptual model of arsenic mobility in shallow alluvial aquifers: case from Bengal Basin. Water Resour Res 55(11):10243–10263

Central Ground Water Board (CGWB) (2025) One in three Punjab groundwater samples exceed uranium limits. The Times of India

Choubisa SL (2023) Fluoride removal from drinking water using Nalgonda technique: a review. J Chil Chem Soc 68(2):5884–5890

Fakhreddine S, Esser BK, Tufenkji N (2024) Limiting mobilization of geogenic contaminants during managed aquifer recharge: a guidance review. Groundw Monit Remediat 44(1):35–48

Fendorf S, Michael HA, van Geen A (2010) Spatial and temporal variations of groundwater arsenic in South and Southeast Asia. Science 328(5982):1123–1127

Hasan MA, Rahman MM, Ahmed KM (2018) Modeling groundwater flow and arsenic transport in arsenic-affected areas of West Bengal and Assam. J Environ Manag 212:453–467

Hussam A (2009) The SONO filter: development, dissemination, and scale-up of a household arsenic removal technology. MIT Press Case Studies in Environmental and Earth Science

Hussam A, Munir AKM (2007) A simple and effective arsenic filter based on composite iron matrix: development and field trials. J Environ Sci Health A 42(12):1869–1878

Ijumulana J, Mrutu A, Mohammed K (2021) Spatial uncertainties in fluoride levels and associated health risks in Northern Tanzania. Groundw Sustain Dev 14:100614

Jing Q, He J, Li Y, Yang X, Peng Y, Wang H et al (2024) Analysis of the spatiotemporal changes in global land cover from 2001 to 2020. Sci Total Environ 908:168354. https://doi.org/10.1016/j.scitotenv.2023.168354

Knappett PSK, Escamilla V, Layton A, McKay LD, Emch M, Williams DE, Sobsey MD (2017) Manganese concentrations in groundwater from glacial and fluvial aquifers of North America and Asia. Sci Total Environ 579:1313–1326

Mahlknecht J, Navarro de León I, Cortés J (2022) Hydrochemical controls on arsenic contamination and its health risks in the Comarca Lagunera region, Mexico. Sci Total Environ 850:158040

Mukherjee A, Bhattacharya P (2022) Sustainability appraisal of arsenic mitigation policies in West Bengal, India. J Infrastruct Policy Dev 10(1)

Mukherjee A, Fryar AE, Rowe HD (2008) Regional hydrostratigraphic and geochemical controls on arsenic contamination in the Bengal Basin: a synthesis. Appl Geochem 23(11):3155–3166

Mukherjee A, Bhattacharya P, Podgorski J (2024) Global arsenic in groundwater: sources, mobilization, and risk implications. Nat Geosci. https://www.nature.com/articles/s43017-024-00519-z

Nickson R, McArthur J, Burgess W, Ahmed KM, Ravenscroft P, Rahman M (1998) Arsenic poisoning of Bangladesh groundwater. Nature 395(6700):338–338

Niu Q, Lu Y, Zhang Z (2022) Removal of iron and manganese from groundwater using modified manganese sand: performance and mechanism. Appl Sci 12(22):11798

Panday DP, Kumar M (2024) Decoding uranium variation over the Indian Peninsula through leveraging standardized precipitation evapotranspiration indices and groundwater level fluctuations. Groundw Sustain Dev:101241. https://doi.org/10.1016/J.GSD.2024.101241

Panday DP, Kumar M, Agarwal V, Torres-Martínez JA, Mahlknecht J (2024) Corroboration of arsenic variation over the Indian Peninsula through standardized precipitation evapotranspiration indices and groundwater level fluctuations: water quantity indicators for water quality prediction. Sci Total Environ:176339. https://doi.org/10.1016/j.scitotenv.2024.176339

Panday DP, Kumari A, Kumar M (2025) Alkalinity-salinity-sustainability: decadal groundwater trends and its impact on agricultural water quality in the Indian Peninsula. Sci Total Environ 978:179459. https://doi.org/10.1016/j.scitotenv.2025.179459

Podgorski J, Berg M (2020) Global threat of arsenic in groundwater. Science 368(6493):845–850

Podgorski J, Berg M (2022) Global analysis and prediction of fluoride in groundwater. Nat Commun 13(1):4232. https://doi.org/10.1038/s41467-022-31920-3

Rahman MM, Hasan MA, Ahmed KM, Ravenscroft P (2015) Subsurface arsenic removal (SAR) technology for drinking water treatment: field trials and modeling. Sci Total Environ 536:160–170

Rango T, Kravchenko J, Atlaw B, McCornick PG, Jeuland M, Merola B, Vengosh A (2014) Groundwater fluoride exposure and human health risk assessment in the Main Ethiopian Rift. Sci Total Environ 496:188–197

Riedel T, Kübeck C (2018) Uranium in groundwater—a synopsis based on a large hydrogeochemical data set. Water Res 129:29–38. https://doi.org/10.1016/j.watres.2017.11.007

Rodríguez-Lado L, Sun G, Berg M, Zhang Q, Xue H, Zheng Q, Johnson CA (2013) Groundwater arsenic contamination throughout China. Science 341(6148):866–868

Shen X, Zhang J, Zhang H, Xu Y, Liu J (2020) Overview of fluoride removal from groundwater using separation techniques. Sep Purif Technol 248:117063

Shukla S, Gupta SK, Singh P (2021) Occurrence and distribution of uranium in groundwater of Punjab, India: a meta-analysis. Sci Total Environ 804:150054

Singh SK, Kumar M, Prasad R (2018) Groundwater arsenic contamination in the Ganga River Basin: a review of distribution, mechanisms, and health risk. Int J Environ Res Public Health 15(8):180

Smedley PL, Kinniburgh DG (2002) A review of the source, behaviour and distribution of arsenic in natural waters. Appl Geochem 17(5):517–568

Smedley PL, Zhang M, Luo Z (2022) Uranium in natural waters and the environment: a global review. Appl Geochem 141:105325

Souza BA, da Silva EM, Pereira JC (2017) Arsenic speciation modeling in natural waters using PHREEQC and Geochemist's Workbench. Braz J Chem Eng 34(3):547–556

State Pollution Control Board (SPCB) Odisha (2025) Dangerously high fluoride levels in Nuapada district groundwater. The Times of India

Thirunavukkarasu OS, Viraraghavan T, Subramanian KS (2003) Arsenic removal from drinking water using granular ferric hydroxide. Water Res 37(19):4359–4366

Van Geen A, Zheng Y, Goodbred S, Horneman A, Aziz Z, Cheng Z, Stute M (2008) Flushing history as a hydrogeological control on regional groundwater arsenic distribution in Bangladesh. Environ Sci Technol 42(7):2283–2288

Yang YS, Park SY, Kim JH (2015) Influence of water chemistry on arsenic mobilization from arsenopyrite under managed aquifer recharge. Environ Sci Technol 49(16):9860–9868

Zhang Q, Sun G, Xue H, Zheng Q (2014) Review of high-arsenic groundwater in Mainland and Taiwan, China. Appl Geochem 41:196–206

Zhang Q, Shen Y, Wu J, Deng Y (2022) Geogenic manganese and iron in groundwater of Southeast Asia and Bangladesh: occurrence, health risk, and mobilization mechanisms. Sci Total Environ 852:158404

Zhang W, Chen W, Zhao L, Liu G (2024) Shifting to biology promotes highly efficient iron removal in groundwater treatment. Water Res 257:121250

Chapter 5
Water Quality Characterization in the Context of Emerging Contaminants

Abstract Emerging contaminants (ECs) have become a primary environmental concern due to their persistence, bioaccumulation potential, and complex interactions with natural matrices. These contaminants, including pharmaceuticals, personal care products, pesticides, volatile organic compounds (VOCs), per- and polyfluoroalkyl substances (PFAS), and micro- and nanoplastics, are increasingly detected in surface water, groundwater, soils, and biota. Unlike conventional pollutants, ECs often occur at trace levels but exhibit high reactivity and synergistic effects, posing risks to ecosystems and human health. This chapter presents a comprehensive overview of the sources, pathways, and characterization approaches of ECs in environmental systems. It draws parallels among diverse contaminant classes, emphasizing the importance of harmonized sampling, extraction, and analytical techniques. Methods such as solid-phase extraction, chromatographic–mass spectrometry (GC-MS, LC-MS/MS), Fourier-transform infrared (FTIR) and Raman spectroscopy, isotopic analysis, and high-resolution mass spectrometry (HRMS) are highlighted as essential for identifying both known and transformation products. Special attention is given to the hydrogeochemical behavior of ECs, including adsorption, complexation, and metal–organic interactions that influence contaminant fate in aquatic and terrestrial environments. The chapter also explores integrating isotopic fingerprinting with machine learning for source attribution and predictive modeling. Finally, it discusses challenges such as analytical standardization, data comparability, and risk assessment, while proposing a multidisciplinary framework for EC monitoring and management. By synthesizing recent advances, including methodologies for characterizing micro- and nanoplastics, this chapter aims to bridge knowledge across contaminant groups and to provide a unified perspective on the analytical, isotopic, and hydrogeochemical characterization of emerging contaminants in the environment.

Keywords Emerging contaminants · Characterization · Spectrometry · Environmental · Isotope

M. Kumar, D. P. Panday, *Enabling Water Security and Sustainability: Global Challenges and Solutions*, SpringerBriefs in Water Science and Technology, https://doi.org/10.1007/978-3-032-21521-5_5

5.1 Introduction

The rapid expansion of industrialization, urbanization, and intensive agriculture has led to the emergence of a new generation of pollutants, collectively termed emerging contaminants (ECs). These substances, ranging from pharmaceuticals, personal care products, and endocrine disruptors to pesticides, microplastics, and nanomaterials, are often unregulated yet increasingly recognized for their ecological and human health implications (Kumar et al. 2021). Their persistence, continuous release, and complex transformation pathways present significant challenges to conventional water-quality assessment and regulatory frameworks. Unlike traditional pollutants, ECs exhibit remarkable chemical diversity and originate from diverse anthropogenic activities, including domestic wastewater, industrial effluents, agricultural runoff, landfill leachate, and atmospheric deposition (Yu et al. 2024). Once introduced into the environment, they interact dynamically with natural matrices through adsorption, complexation, and redox-mediated transformations. Their widespread detection in surface and groundwater systems is particularly concerning due to their mobility, persistence, and potential for long-range transport (Tripathi et al. 2023).

Comprehensive characterization of ECs is essential to understand their sources, distribution, and environmental behavior. Characterization typically involves sampling, extraction, identification, quantification, and risk evaluation. The development of standardized analytical methodologies is crucial for ensuring data comparability and reproducibility across studies. Advanced techniques such as Fourier-transform infrared (FTIR) and Raman spectroscopy, scanning electron microscopy (SEM) (Halko et al. 2024), and chromatographic systems coupled with mass spectrometry (GC–MS, LC–MS/MS) enable sensitive and accurate detection of organic and polymeric contaminants. In addition, compound-specific isotope analysis (CSIA) provides valuable insights into contaminant sources and transformation mechanisms. At the same time, high-resolution mass spectrometry (HRMS) supports non-target and suspect screening of previously unidentified compounds (Meher and Zarouri 2025). This chapter synthesizes current understanding of the occurrence, behavior, and analytical characterization of emerging contaminants, emphasizing the need for harmonized protocols, cross-contaminant comparability, and the integration of advanced analytical tools to support sustainable monitoring and management of environmental pollution.

5.2 Sources and Pathways of Emerging Contaminants

ECs originate from diverse anthropogenic activities and enter natural systems through both point and non-point sources (Namieśnik et al. 2014). Their distribution in the environment is governed by factors such as production and consumption patterns, physicochemical properties, degradation kinetics, and environmental persistence. Understanding these sources and pathways is essential for tracing contaminant fluxes, evaluating their long-term fate, and formulating mitigation strategies.

5.2.1 Primary and Secondary Sources

EC sources are broadly categorized into primary, referring to direct environmental releases, and secondary, representing derivatives or transformation products. This distinction, widely recognized in ecological pollution studies, separates contaminants introduced directly into the environment from those generated through degradation or transformation processes.

Municipal wastewater effluents are among the most significant primary sources, containing complex mixtures of pharmaceuticals, personal care products, endocrine-disrupting compounds, and microplastics derived from domestic use (Ahirvar et al. 2025). Industrial discharges contribute synthetic solvents, dyes, surfactants, and per- and polyfluoroalkyl substances (PFAS) used in coatings, lubricants, and packaging materials (Dogra et al. 2024). Agricultural practices release pesticides, herbicides, and veterinary antibiotics through irrigation runoff and leaching, while animal husbandry operations introduce hormones, antibiotics, and dissolved organic carbon via manure and wastewater management systems. Urban runoff also acts as a diffuse source, transporting hydrocarbons, tire wear particles, and heavy metals from roads and construction sites into nearby aquatic systems.

Secondary ECs arise from physicochemical transformation or biological degradation of primary pollutants (Namieśnik et al. 2014). Photodegradation of plastics, pesticides, and pharmaceuticals in surface waters, microbial transformation producing metabolites with altered toxicological properties, and chemical oxidation or hydrolysis forming halogenated by-products such as trihalomethanes (THMs) in treated water exemplify secondary formation processes. Differentiating between primary and secondary origins is critical for source tracking and risk assessment, as transformation products often exhibit greater persistence or bioavailability than their parent compounds.

5.2.2 Environmental Compartments and Pathways

Once released, ECs are distributed among four main environmental compartments: surface water, groundwater, soil/sediment, and atmosphere through multiple interlinked pathways (Fig. 5.1).

Surface water bodies, such as rivers, lakes, and wetlands, act as both sinks and transport media for ECs. Effluents from wastewater treatment plants (WWTPs) remain a dominant input route, often containing residual pharmaceuticals, PFAS,

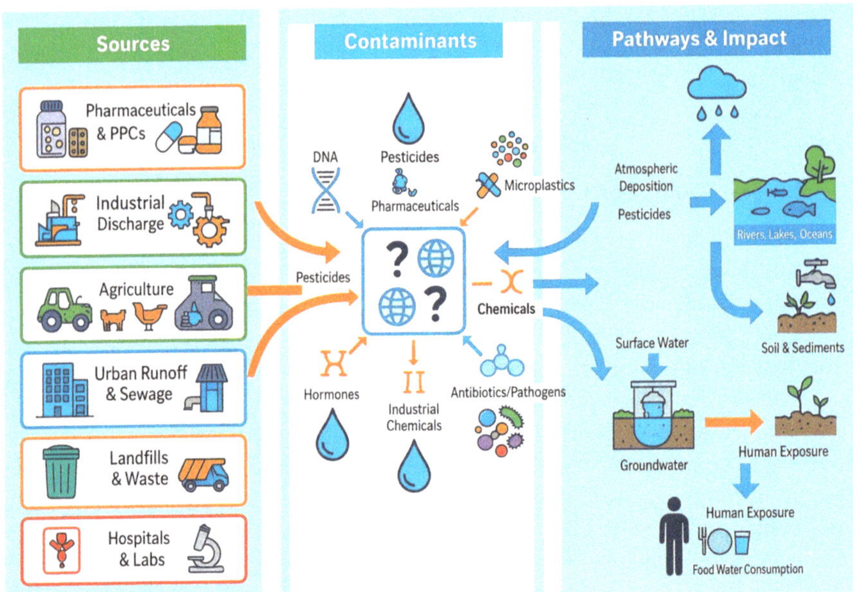

Fig. 5.1 Sources, contaminants, and environmental pathways of emerging pollutants. Emerging contaminants originate from a diverse range of anthropogenic activities, including industrial discharge, agricultural runoff, landfill leachate, urban wastewater, and hospital effluents. These sources release numerous chemical and biological pollutants, such as pharmaceuticals, personal care products, pesticides, hormones, antibiotics, and microplastics. Once introduced into the environment, contaminants are transported through atmospheric deposition, surface water, and infiltration processes, ultimately reaching rivers, lakes, soil, and groundwater systems

and microplastics even after tertiary treatment. The spatial and temporal distribution of these contaminants is influenced by hydrodynamic processes, sediment–water interactions, and seasonal variations that regulate dilution, resuspension, and photodegradation (Panday et al. 2025).

Groundwater contamination occurs through leaching, infiltration, and hydraulic connectivity between surface and subsurface systems. Similar to the mechanisms observed for micro- and nanoplastics, dissolved or colloid-associated ECs migrate according to soil permeability, organic carbon content, and prevailing redox gradients. Persistent species such as PFAS, nitrates, and certain pharmaceuticals can migrate over long distances and accumulate in confined aquifers, posing significant threats to drinking-water security (Glassmeyer et al. 2023).

Soil and sediment act as both reservoirs and transformation media for ECs. Adsorption onto clay minerals, organic matter, and iron oxides may temporarily immobilize contaminants, while microbial activity can induce biotransformation (Neculita et al. 2020). Agricultural soils irrigated with reclaimed wastewater or amended with biosolids often exhibit elevated concentrations of antibiotics, hormones, and plastic fragments, contributing to long-term soil contamination. The dynamic exchange between sediment and overlying water facilitates the continuous cycling of ECs, influencing their persistence and ecological exposure potential.

Atmospheric transport plays a key role in the long-range dispersion of volatile or particle-bound contaminants. Pesticides, volatile organic compounds (VOCs), and microplastics can be emitted through volatilization, abrasion, or combustion processes. Airborne particles can later deposit on terrestrial or aquatic surfaces, providing a secondary input route to soil and water systems.

5.2.3 *Fate, Transport, and Transformation*

The fate of emerging contaminants in the environment is governed by their intrinsic physicochemical characteristics, such as hydrophobicity, vapor pressure, solubility, and susceptibility to degradation, as well as by external environmental conditions, including temperature, redox potential, and biological activity (Ghosh et al. 2024). Following their release, ECs may partition between aqueous, solid, and gaseous phases through adsorption onto suspended solids, organic colloids, or microplastic particles, processes that significantly influence residence time, transport, and bioavailability (Panday and Kumar 2024; Kumar et al. 2025).

Both abiotic and biotic transformations contribute to altering ECs in the environment. Abiotic processes, including photolysis, oxidation, and hydrolysis, lead to chemical modification or breakdown, whereas biotic mechanisms such as enzymatic degradation and microbial metabolism produce intermediate products that can vary in toxicity. These transformations, combined with sorption–desorption dynamics in soils and sediments, determine the mobility and environmental persistence of ECs. Interactions with co-contaminants, such as dissolved organic carbon, trace metals, and reactive oxygen species, can further modify mobility, complexation behavior, and ecological impacts.

In aquatic and terrestrial food webs, ECs may bioaccumulate in organisms and transfer across trophic levels, as demonstrated for pharmaceuticals, heavy metals, and microplastics. The combined influence of partitioning, transformation, and bioaccumulation results in a wide range of environmental half-lives and complex mixtures of transformation products, many of which remain analytically unresolved. Comprehensive evaluation of these processes, therefore, requires the integrated application of chemical, isotopic, and spectroscopic tools to elucidate contaminant pathways and transformation mechanisms.

5.3 Hydrochemical Interactions

The environmental behavior of ECs is governed by the complex interplay between their intrinsic physicochemical properties and the hydrogeochemical characteristics of the surrounding environment. Once released, these compounds undergo processes such as sorption, complexation, transformation, and migration, which collectively determine their persistence, bioavailability, and ecological impact. Parameters including pH, redox potential, ionic strength, and organic carbon content exert dominant control over these processes, particularly in dynamic aquatic and subsurface systems where strong gradients in these factors exist (Daughton and Ternes 1999; Richardson 2010).

The mobility of ECs is primarily dictated by their partitioning between aqueous and solid phases. Hydrophobic organic contaminants such as pesticides, PFAS, and hydrocarbons preferentially associate with organic matter and mineral surfaces. In contrast, polar and ionizable species, including many pharmaceuticals and nutrients, remain mainly in the dissolved phase (Masinga et al. 2024). The pH of the medium controls the ionization state of these compounds and thereby influences their adsorption affinity. Under alkaline conditions, acidic pharmaceuticals predominantly exist in anionic forms with low sorption potential.

In contrast, under acidic conditions, basic compounds acquire positive charges, enhancing their attachment to negatively charged mineral and organic surfaces. Redox potential further regulates the speciation and solubility of both metals and redox-sensitive organics; reducing environments favor the mobilization of metals such as manganese and iron, whereas oxidizing conditions promote their precipitation as oxides or hydroxides. Variations in ionic strength modify electrostatic interactions, affecting the aggregation and desorption behavior of colloidal and dissolved species (Linnik et al. 2023).

Stable isotope analysis provides a powerful approach to tracing contaminant sources and transformation processes. Variations in the isotopic ratios of carbon ($\delta^{13}C$), nitrogen ($\delta^{15}N$), hydrogen ($\delta^{2}H$), and sulfur ($\delta^{34}S$) yield valuable insights into contaminant origin, extent of degradation, and reaction pathways. Compound-specific isotope analysis (CSIA) quantifies isotope fractionation associated with chemical and biological reactions, allowing direct evaluation of degradation kinetics. When integrated with hydrochemical and spectroscopic datasets, isotopic methods form a robust multi-proxy framework capable of distinguishing anthropogenic inputs, natural background processes, and transformation pathways within complex aquifer systems (Elsner and Imfeld 2016; Kendall and Aravena 2000).

The collective influence of hydrochemical factors determines the environmental fate of emerging contaminants. Within natural aquatic and subsurface systems, variations in pH, redox potential, and organic carbon content regulate adsorption, complexation, and redox-mediated transformations, ultimately controlling contaminant mobility and persistence. Microbial activity further contributes to chemical evolution through biodegradation and isotopic fractionation, reflecting ongoing redox and transformation processes within the system. Understanding these coupled interactions provides a mechanistic basis for evaluating contaminant transport and persistence under varying geochemical conditions. This integrated perspective underscores that the fate of emerging contaminants cannot be understood independently of the hydrogeochemical environment, in which organic matter, mineral phases, and redox gradients collectively dictate contaminant behavior and ecological risk.

5.4 Characterization and Analytical Techniques

The characterization of ECs requires an integrated analytical framework combining systematic sampling, optimized extraction and preconcentration, and high-resolution instrumental techniques (Domini et al. 2005). Together, these processes allow both qualitative and quantitative assessment of contaminants across diverse environmental matrices. Because ECs often occur at trace or ultra-trace concentrations, their reliable identification demands exceptional analytical precision, rigorous contamination control, and methodological standardization. Despite significant advances in recent decades, the absence of universally accepted protocols continues to hinder interstudy comparability (Ścigalski and Kosobucki 2020). The analytical philosophy underpinning EC characterization increasingly emphasizes a holistic approach in which sampling, extraction, and instrumental determination are interdependent components of a single workflow. As summarized in Table 5.1, various analytical and characterization techniques differ significantly in their sensitivity, selectivity, and applicability to emerging contaminants. Figure 5.2 highlights the evolution of the characterisation of emerging contaminants.

This figure summarizes how analytical tools, catalysts, and contaminant focus have shifted from targeted monitoring of specific compounds to advanced, high-resolution molecular forensics for complex contaminant mixtures.

5.4.1 *Sampling Strategies*

Sampling represents the foundation of contaminant characterization, as it determines the accuracy, representativeness, and reproducibility of all subsequent analyses. In surface water and groundwater systems, samples are collected using pre-cleaned glass or metal containers to minimize contamination. At the same time, temperature and redox conditions are maintained to preserve in situ chemistry. Composite or depth-integrated sampling is often employed to capture temporal and spatial variability, particularly in systems influenced by seasonal hydrological changes or anthropogenic discharge. Sediment and soil sampling, crucial for understanding contaminant storage and transformation, typically employs coring or grab techniques designed to minimize stratigraphic disturbance. The exclusive use of non-plastic materials in sampling equipment is critical, particularly for hydrophobic and polymeric contaminants. Following collection, samples are preserved under controlled light and temperature conditions to prevent volatilization, oxidation, or microbial alteration before analysis.

Table 5.1 Comparative overview of primary analytical and characterization techniques for emerging contaminants

Technique/instrument	Sample matrices/phase	Sample preparation and extraction method	Target contaminant	Analytical capability/detection range	Required sample volume/mass	Advantages	Limitations/challenges	References
Gas Chromatography–Mass Spectrometry (GC-MS/GC-MS/MS)	Water, air, sediment, sludge; dissolved or volatile phase	Solvent extraction or liquid–liquid extraction; derivatization for polar analytes	Volatile and semi-volatile organic compounds (VOCs, PAHs, pesticides)	1–100 ng/L	50–500 mL od 1–10 g	Established libraries are suitable for volatile ECs	Requires derivatization for polar compounds; limited to thermally stable, volatile analytes; possible matrix interference	Meher and Zarouri (2025)
Liquid Chromatography–Tandem Mass Spectrometry (LC–MS/MS)	Water, soil, sludge, biota; dissolved or colloidal phase	Solid-phase extraction (SPE)	Pharmaceuticals, PFAS, hormones, polar pesticides, and personal care products	0.1–10 ng/L	250 mL –2 L	High sensitivity and selectivity, quantitative	Matrix suppression and ionization variability; high operational cost; limited spectral libraries	Čelić et al. (2017)

Technique/ instrument	Sample matrices/ phase	Sample preparation and extraction method	Target contaminant	Analytical capability/ detection range	Required sample volume/mass	Advantages	Limitations/ challenges	References
Liquid Chromatography-High-Resolution Mass Spectrometry (LC-HRMS) (Orbitrap, TOF)	Water, wastewater, sediment; dissolved phase	SPE or direct injection after filtration	Unknown and transformation products; non-target and suspect screening	0.1–10 ng/L	100 mL –1 L	Non-targeted screening, high resolution	Complex data, time-intensive analysis; lack of standardized databases	Čelić et al. (2017)
Fourier Transform Infrared (FTIR) Spectroscopy	Sediments, suspended solids, microplastics; solid phase	Filtration, drying, and pressing onto KBr discs or reflective substrates	Polymeric materials, functionalized organic residues	Detects characteristic IR bands; particle size ≥10 μm	Minimal (solid/ plastic fragment)	Nondestructive, rapid identification of polymers and functional groups	Limited sensitivity for very small particles; spectral overlap with natural organic matter	Quintelas et al. (2020)
Raman Spectroscopy	Solids, microplastics, sediments, biofilms	Filtration and particle isolation on glass or metal substrates	Polymeric, carbonaceous, or pigment-bearing contaminants	High spatial resolution (≤1 μm); qualitative molecular ID	Minimal (solid/ plastic fragment)	Minimal sample preparation, suitable for colored polymers	Fluorescence interference; longer acquisition time	Almaviva et al. (2022)

(continued)

Table 5.1 (continued)

Technique/instrument	Sample matrices/phase	Sample preparation and extraction method	Target contaminant	Analytical capability/detection range	Required sample volume/mass	Advantages	Limitations/challenges	References
Inductively Coupled Plasma–Mass Spectrometry (ICP–MS)	Water, soil, sediment, sludge, biota; dissolved or particulate phase	Acid digestion (HNO_3–HCl, aqua regia) and dilution	Trace metals, metalloids, metal–organic complexes	Sub-ppb detection; multi-element quantification	10–50 mL/0.1–1 g	Ultra-trace sensitivity; simultaneous multi-element detection; rapid throughput	Requires digestion; prone to matrix interference; expensive maintenance	Shyamalagowri et al. (2023)
X-Ray Photoelectron Spectroscopy (XPS)	Solids, mineral coatings, polymers; surface phase	Vacuum drying; mounted on conductive stubs	Metal oxides, functionalized polymers, surface films	Surface-sensitive (10 nm depth); oxidation state and bonding environment	1–20 g	Provides detailed surface chemistry; practical for metal binding and oxidation studies	High vacuum requirement; limited sample size; non-representative for bulk composition	Richardson (2008)
Pyrolysis–Gas Chromatography–Mass Spectrometry (Py–GC–MS)	Plastics, sediments, and atmospheric dust	Direct pyrolysis of solids under an inert atmosphere	Polymeric materials, additives, and organic residues	Identification via pyrolytic fragment spectra; detection limit ~$\mu g\ g^{-1}$	1–20 g	No solvent extraction needed; highly specific for polymers	Destructive; quantification difficult; limited to organic solids	Picó and Barceló (2020)

Technique/ instrument	Sample matrices/ phase	Sample preparation and extraction method	Target contaminant	Analytical capability/ detection range	Required sample volume/mass	Advantages	Limitations/ challenges	References
Fluorescence and UV–visible Spectroscopy	Water, wastewater, dissolved organic fraction	Filtration and dilution; excitation–emission matrix (EEM)	Dissolved organic carbon, aromatic, and fluorescent ECs	Qualitative and semi-quantitative; ng L^{-1}–µg L^{-1} range	Minimal (solid/ plastic fragment)	Low cost, high throughput; rapid screening of DOC and aromatic compounds	Poor molecular resolution; sensitive to pH, turbidity, and quenching	Richardson (2008)
Compound-Specific Isotope Analysis (CSIA)	Groundwater, soil, organic extracts; dissolved phase	Extraction and purification of target compounds; GC or LC interface to isotope ratio MS	Isotopically traceable ECs (VOCs, nitrates, solvents, pesticides)	$\delta^{13}C$, $\delta^{15}N$, $\delta^{2}H$, $\delta^{34}S$ precision ±0.5‰; compound-level isotope quantification	10–50 mL / 0.1–1 g	Identifies sources and transformation processes; distinguishes biodegradation from dilution.	Requires relatively high analyte concentration; technically demanding; limited availability of compound-specific standards	Cui et al. (2021)

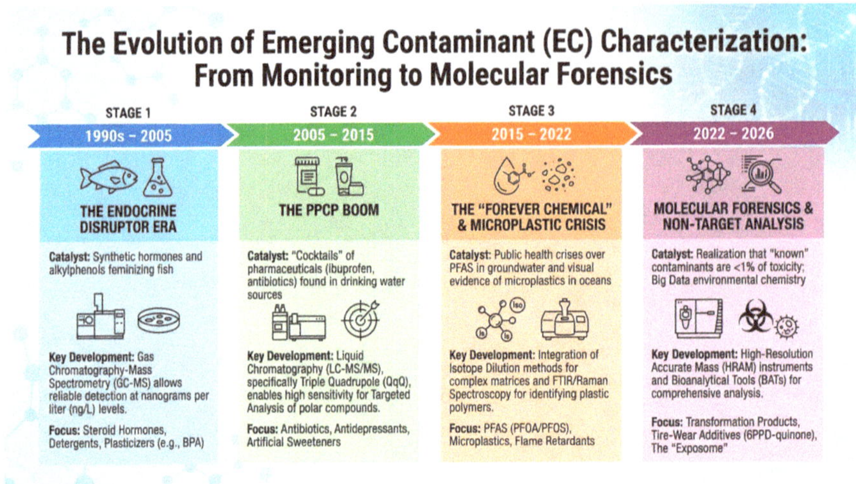

Fig. 5.2 Evolution of emerging contaminant (EC) characterization from monitoring to molecular forensics, highlighting four stages: the endocrine disruptor era (1990s–2005), the PPCP boom (2005–2015), the "forever chemical" and microplastic crisis (2015–2022), and the current phase of molecular forensics and non-target analysis (2022–2026)

5.4.2 *Extraction and Pre-treatment Methods*

Pre-treatment and extraction steps are central to isolating contaminants from complex matrices and concentrating analytes to detectable levels. In aqueous samples, solid-phase extraction (SPE) remains the most widely used approach for enriching pharmaceuticals, pesticides, and endocrine-disrupting compounds via adsorption onto hydrophobic or ion-exchange sorbents. Liquid–liquid extraction (LLE) and purge-and-trap techniques are preferred for volatile organics due to their efficiency in recovering low-molecular-weight compounds (Ścigalski and Kosobucki 2020).

Metal analysis generally involves acid digestion, which dissolves suspended solids and releases bound metals for quantification. For polymeric materials, such as micro- and nanoplastics, filtration and density separation techniques are used to isolate particles based on size and density. Mahapatra et al. (2024) demonstrated that sequential filtration followed by chemical or enzymatic digestion of organic matter yields high recovery of plastic particles from water and sediment samples, an approach now being adapted to improve the efficiency of organic contaminant extraction. Throughout all pre-treatment stages, the inclusion of procedural blanks, duplicates, and field controls is essential to monitor contamination and ensure analytical reliability.

5.4.3 Chromatographic and Mass Spectrometric Analysis

Chromatographic–mass spectrometric techniques form the cornerstone of EC characterization, providing unparalleled molecular specificity and detection sensitivity. Gas chromatography–mass spectrometry (GC–MS) and liquid chromatography–tandem mass spectrometry (LC–MS/MS) remain the primary tools for targeted analysis of known organic contaminants, enabling detection of pharmaceuticals, pesticides, and personal care products at nanogram-per-liter levels (Ranjan et al. 2023). The advent of high-resolution mass spectrometry (HRMS), including Orbitrap and time-of-flight (ToF) instruments, has transformed environmental analysis by enabling non-target analysis (NTA) (Manz et al. 2023). These techniques identify unknown compounds and transformation products based on accurate mass, isotopic patterns, and fragmentation profiles. NTA approaches are especially valuable for complex environmental matrices that contain thousands of organic features, of which only a fraction correspond to known contaminants (Manz et al. 2023).

5.4.4 Spectroscopic and Microscopic Techniques

Spectroscopic techniques complement chromatographic analysis by providing molecular, structural, and functional information. Fourier-transform infrared (FTIR) and Raman spectroscopy, widely used for microplastic identification, are equally effective at detecting specific functional groups in organic residues and degradation products (Xu et al. 2019). FTIR identifies vibrational modes associated with characteristic bonds, while Raman spectroscopy provides high spatial resolution and sensitivity to aromatic or crystalline structures.

Microscopic methods such as scanning electron microscopy (SEM) and transmission electron microscopy (TEM) extend the analytical range by visualizing morphological features at micro- to nanometer scales (Inkson 2016). When coupled with energy-dispersive X-ray spectroscopy (EDS), these instruments provide elemental mapping that reveals contaminant–mineral interactions and surface composition. Such morphological and compositional characterization, as demonstrated by Mahapatra et al. (2024), is increasingly applied to particulate organic residues and colloidal metal–organic complexes (Quintelas et al. 2020).

5.4.5 Elemental and Surface Analytical Techniques

The characterization of metals and metalloids associated with ECs is achieved through inductively coupled plasma mass spectrometry (ICP–MS) and X-ray photoelectron spectroscopy (XPS). ICP–MS provides rapid, multi-element quantification with sub-part-per-billion sensitivity, whereas XPS delivers surface-specific information on oxidation states and bonding environments (Isaacs et al. 2021). Together, these techniques elucidate mechanisms of metal binding on microplastic and organic surfaces, including electrostatic attraction, hydrogen bonding, and surface complexation (Richardson 2010). Their combined application extends naturally to groundwater systems, where complexation between metals and dissolved organic matter influences contaminant mobility and transformation.

5.4.6 *Thermal and Pyrolytic Techniques*

Thermal and pyrolytic methods broaden the analytical window by identifying polymeric or carbonaceous materials based on their thermal behavior. Thermogravimetric analysis (TGA) and differential scanning calorimetry (DSC) distinguish materials by thermal stability and decomposition patterns. At the same time, pyrolysis–gas chromatography–mass spectrometry (Py–GC–MS) enables direct identification of polymer fragments and additives (Zhu et al. 2004). Although inherently destructive, these methods are invaluable for characterizing microplastics, rubber particles, and thermally resistant residues. For organic pollutants, thermochemical degradation before GC–MS analysis provides diagnostic fragmentation profiles that aid in compound identification and quantification (Vaye et al. 2022). Figure 5.3 compares analytical techniques based on detection sensitivity and sample volume requirements using linear scales to emphasize practical monitoring constraints. High-sensitivity mass spectrometric methods require larger sample volumes, whereas spectroscopic techniques operate with minimal sample input but lower sensitivity.

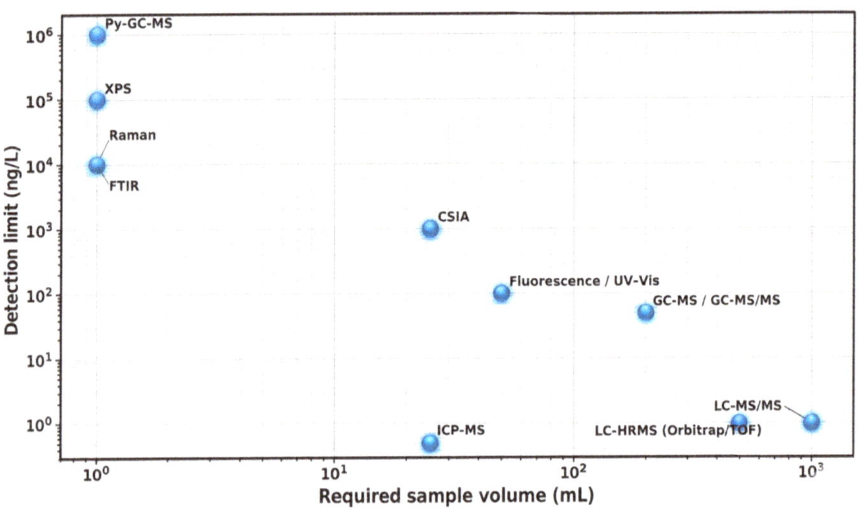

Fig. 5.3 Scatter plot comparing analytical techniques used for emerging contaminant analysis based on detection limits and required sample volumes. Labels are shown for each technique to illustrate trade-offs between analytical sensitivity and practical sampling constraints

5.4.7 *Toward an Integrated Analytical Framework*

These analytical approaches together form a multi-tiered characterization framework that captures the chemical, structural, and isotopic diversity of emerging contaminants. From sampling and extraction to chromatographic quantification and isotopic tracing, each step provides unique yet complementary information on contaminant identity, concentration, and transformation. The convergence of chromatographic precision, spectroscopic structural elucidation, elemental quantification, and isotopic analysis has transformed environmental chemistry into a discipline capable of addressing the complexity of modern pollution. As emphasized by Schwesig et al. (2011), developing standardized and harmonized methodologies remains central to achieving global comparability in contaminant assessment. Extending this integrated analytical philosophy to all classes of ECs ensures that environmental research can not only detect contaminants but also interpret and predict their behavior in evolving ecological systems.

Despite substantial analytical progress, the comprehensive characterization of ECs across environmental compartments remains an enduring challenge. The chemical diversity of ECs, their ultra-trace concentrations, and the complexity of ecological matrices hinder accurate detection and interpretation. Current monitoring frameworks often target a limited set of compounds, leaving most contaminants uncharacterized. Standardized methods, certified reference materials, and interlaboratory comparability are still lacking, which restricts the generation of coherent global datasets. The field must now evolve from descriptive analysis to integrative and predictive understanding. Combining non-target and suspect screening with isotopic approaches such as CSIA can link contaminant identity, origin, and transformation pathways within a single framework. Machine learning and chemometric models can further classify contaminants and predict environmental behavior, while reactive transport modeling can simulate their mobility and degradation in natural systems.

5.5 Conclusion

The characterization of emerging contaminants has progressed from a narrow analytical pursuit to a multidisciplinary scientific framework central to environmental research. As anthropogenic activities continue to diversify, the range of pollutants entering natural systems expands, encompassing pharmaceuticals, pesticides, PFAS, and polymeric residues such as micro- and nanoplastics. Despite their chemical diversity, these contaminants share persistent, reactive, and potentially toxic behavior even at trace levels. Understanding their fate requires an integrated approach that combines hydrogeochemical insight with advanced analytical and isotopic techniques.

This chapter has underscored how EC characterization depends on the interplay between environmental processes and analytical methodologies. Hydrogeochemical parameters such as pH, redox potential, and organic carbon control contaminant speciation and mobility, while chromatographic, spectroscopic, and isotopic tools provide the precision necessary for molecular identification and transformation assessment. However, the absence of standardized protocols and global comparability continues to limit reliable cross-system analysis.

Looking forward, characterization must bridge chemistry, hydrogeology, and data science. The fusion of high-resolution analytical data with isotopic tracing and machine learning will allow the transition from empirical observation to predictive capability. Through methodological harmonization, international collaboration, and open-access data integration, the scientific community can move toward a future where emerging contaminants are no longer elusive but systematically identified, quantified, and interpreted within the global biogeochemical cycle. Such progress represents not only an analytical achievement but also a prerequisite for sustainable management of environmental and water resources.

References

Ahirvar BP, Das P, Kumar R, Panday DP, Das N, Srivastva PK, Kumar M (2025) Weathering in the subsurface environment of the Gondwana deposits intensified by the cumulative impacts of mining and agriculture practices. J Contam Hydrol:104630. https://doi.org/10.1016/j.jconhyd.2025.104630

Almaviva S, Artuso F, Giardina I, Lai A, Pasquo A (2022) Fast detection of different water contaminants by Raman spectroscopy and surface-enhanced Raman spectroscopy. Sensors 22(21):8338

Čelić M, Insa S, Škrbić B, Petrović M (2017) Development of a sensitive and robust online dual column liquid chromatography-tandem mass spectrometry method for the analysis of natural and synthetic estrogens and their conjugates in river water and wastewater. Analytical and Bioanalytical Chemistry 409(23):5427–5440

Cui G, Lartey-Young G, Chen C, Ma L (2021) Photodegradation of pesticides using compound-specific isotope analysis (CSIA): a review. RSC Adv 11(41):25122–25140

Daughton CG, Ternes TA (1999) Pharmaceuticals and personal care products in the environment: agents of subtle change? Environ Health Perspect 107(Suppl 6):907

Dogra K, Lalwani D, Dogra S, Panday DP, Raval N, Trivedi M et al (2024) Indian and global scenarios of Bisphenols A distribution and its new analogues: prevalence & probability exceedance. J Hazard Mater:135128. https://doi.org/10.1016/j.jhazmat.2024.135128

Domini CE, Hristozov D, Almagro B, Román IP, Prats S, Canals A (2005) Sample preparation for chromatographic analysis of environmental samples. In: Chromatographic analysis of the environment, Chromatographic science series, vol 93. Taylor & Francis, pp 31–132

Elsner M, Imfeld G (2016) Compound-specific isotope analysis (CSIA) of micropollutants in the environment—current developments and future challenges. Curr Opin Biotechnol 41:60–72

Ghosh R, Parde D, Bhaduri S, Rajpurohit P, Behera M (2024) Occurrence, fate, transport, and removal technologies of emerging contaminants: a review on recent advances and future perspectives. CLEAN–Soil Air Water 52(12):2300259

Glassmeyer ST, Burns EE, Focazio MJ, Furlong ET, Gribble MO, Jahne MA et al (2023) Water, water everywhere, but every drop unique: challenges in the science to understand the role of contaminants of emerging concern in the management of drinking water supplies. GeoHealth 7(12):e2022GH000716

Halko R, Pavelek D, Kaykhaii M (2024) High performance liquid chromatography-Fourier transform infrared spectroscopy coupling: a comprehensive review. Crit Rev Anal Chem 56:1–12

Inkson BJ (2016) Scanning electron microscopy (SEM) and transmission electron microscopy (TEM) for materials characterization. In: Materials characterization using nondestructive evaluation (NDE) methods. Woodhead Publishing, pp 17–43

Isaacs MA, Davies-Jones J, Davies PR, Guan S, Lee R, Morgan DJ, Palgrave R (2021) Advanced XPS characterization: XPS-based multi-technique analyses for comprehensive understanding of functional materials. Mater Chem Front 5(22):7931–7963

Kendall C, Aravena R (2000) Nitrate isotopes in groundwater systems. In: Environmental tracers in subsurface hydrology. Springer US, Boston, pp 261–297

Kumar M, Vithanage M, Mohapatra S, Acharya K (eds) (2021) Legacy, pathogenic and emerging contaminants in the environment. CRC Press

Kumar M, Panday DP, Goswami R, Neog N, Bahukhandi KD, Zhang J, Bhattacharya P (2025) Dynamics of arsenic co-contamination in the groundwater of the Eastern Himalayas: the biogeochemical drivers and GRACE-based evidence. ACS EST Water. https://doi.org/10.1021/acsestwater.4c00106

Linnik P, Osadchyi V, Osadcha N, Linnik R (2023) Redox potential as an important characteristic of the chemical and biological state of surface waters. Chem Ecol 39(6):640–672

Mahapatra S, Maity JP, Singha S, Mishra T, Dey G, Samal AC, Bhattacharya P (2024) Microplastics and nanoplastics in environment: Sampling, characterization and analytical methods. Groundwater for Sustainable Development 26:101267

Manz KE, Feerick A, Braun JM, Feng YL, Hall A, Koelmel J et al (2023) Non-targeted analysis (NTA) and suspect screening analysis (SSA): a review of examining the chemical exposome. J Expo Sci Environ Epidemiol 33(4):524–536

Masinga P, Simbanegavi TT, Makuvara Z, Marumure J, Chaukura N, Gwenzi W (2024) Emerging organic contaminants in the soil–plant-receptor continuum: transport, fate, health risks, and removal mechanisms. Environ Monit Assess 196(4):367

Meher AK, Zarouri A (2025) Environmental applications of mass spectrometry for emerging contaminants. Molecules 30(2):364

Namieśnik J, Wolska L, Czernych R, Gałęzowska G, Cieszyńska M (2014) Classifying the transformation products (TPs) of emerging contaminants (ECs) for prioritizing research into their impact on the environment and human health. In: Transformation products of emerging contaminants in the environment. Wiley, pp 1–64

Neculita CM, Coudert L, Rosa E, Mulligan CN (2020) Future prospects for treating contaminants of emerging concern in water and soils/sediments. In: Advanced nano-bio technologies for water and soil treatment. Springer International Publishing, Cham, pp 589–605

Panday DP, Kumar M (2024) Decoding uranium variation over the Indian Peninsula through leveraging standardized precipitation evapotranspiration indices and groundwater level fluctuations. Groundw Sustain Dev:101241. https://doi.org/10.1016/J.GSD.2024.101241

Panday DP, Kumari A, Kumar M (2025) Alkalinity-salinity-sustainability: decadal groundwater trends and its impact on agricultural water quality in the Indian Peninsula. Sci Total Environ 978:179459. https://doi.org/10.1016/j.scitotenv.2025.179459

Picó Y, Barceló D (2020) Pyrolysis gas chromatography-mass spectrometry in environmental analysis: focus on organic matter and microplastics. TrAC Trends Anal Chem 130:115964

Quintelas C, Melo A, Costa M, Mesquita DP, Ferreira EC, Amaral AL (2020) Environmentally-friendly technology for rapid identification and quantification of emerging pollutants from wastewater using infrared spectroscopy. Environ Toxicol Pharmacol 80:103458

Ranjan S, Chaitali ROY, Sinha SK (2023) Gas chromatography–mass spectrometry (GC-MS): a comprehensive review of synergistic combinations and their applications in the past two decades. J Anal Sci Appl Biotechnol 5(2):72–85

Richardson SD (2008) Environmental mass spectrometry: emerging contaminants and current issues. Analytical chemistry 80(12):4373–4402
Richardson SD (2010) Environmental mass spectrometry: emerging contaminants and current issues. Anal Chem 82(12):4742–4774
Schwesig D, Borchers U, Chancerelle L, Dulio V, Eriksson U, Farré M et al (2011) A harmonized European framework for method validation to support research on emerging pollutants. TrAC Trends Anal Chem 30(8):1233–1242
Ścigalski P, Kosobucki P (2020) Recent materials developed for dispersive solid phase extraction. Molecules 25(21):4869
Shyamalagowri S, et al. (2023) Occurrence and fate of emerging contaminants in water bodies: A review on the application of ICP-MS and other analytical techniques
Tripathi S, Kumar R, Mondal P, Thakur AK, Singh S, Kaur G et al (2023) Urban fingerprinting on the River Ganges: a perspective of emerging contaminants status. Total Environ Res Themes 5:100028
Vaye O, Ngumbu RS, Xia D (2022) A review of the application of comprehensive two-dimensional gas chromatography MS-based techniques for the analysis of persistent organic pollutants and ultra-trace level of organic pollutants in environmental samples. Rev Anal Chem 41(1):63–73
Xu JL, Thomas KV, Luo Z, Gowen AA (2019) FTIR and Raman imaging for microplastics analysis: state of the art, challenges and prospects. TrAC Trends Anal Chem 119:115629
Yu Y, Wang S, Yu P, Wang D, Hu B, Zheng P, Zhang M (2024) A bibliometric analysis of emerging contaminants (ECs) (2001–2021): evolution of hotspots and research trends. Sci Total Environ 907:168116
Zhu P, Sui S, Wang B, Sun K, Sun G (2004) A study of pyrolysis and pyrolysis products of flame-retardant cotton fabrics by DSC, TGA, and PY–GC–MS. J Anal Appl Pyrolysis 71(2):645–655

Chapter 6
One Health: Need for Water Quality

Abstract This chapter positions water as a central connector of human, animal, and ecosystem health and argues that water quality must be understood and managed through a One Health lens. The study outlines how the physical, chemical, microbial, and ecological dimensions of water quality jointly determine exposure to pathogens, chemical pollutants, and antimicrobial resistance across shared human–animal–environmental interfaces. Drawing on examples including cholera outbreaks in South Asia and Haiti, Cryptosporidium transmission from livestock to communities in Kenya and Uruguay, pesticide-contaminated drinking and irrigation water in Ethiopian highlands, and cyanobacterial toxin–driven mortality of elephants in southern Africa, we illustrate how urbanization, agricultural intensification, industrial effluents, climate-driven extremes and untreated wastewater act as interconnected drivers of waterborne disease, antimicrobial resistance and biodiversity loss. Recent advances in monitoring—from traditional culture-based methods to molecular diagnostics, metagenomics, multi-omics and wastewater-based epidemiology—show how these tools reveal a far richer picture of microbial and chemical hazards than conventional surveillance, while remaining dependent on governance, infrastructure and equity considerations for impact. Reframing water quality within a One Health paradigm and embedding advanced monitoring in integrated, cross-sectoral management are essential for reducing infectious disease burdens, slowing the spread of antimicrobial resistance, protecting food and livelihood security, and strengthening socio-ecological resilience under accelerating Anthropocene change.

Keywords One Health · Water quality · Antimicrobial resistance · Wastewater-based epidemiology · Climate change

M. Kumar, D. P. Panday, *Enabling Water Security and Sustainability: Global Challenges and Solutions*, SpringerBriefs in Water Science and Technology,
https://doi.org/10.1007/978-3-032-21521-5_6

6.1 Introduction

6.1.1 Concept of One Health *and Its Relevance to Water*

The One Health philosophy is an interprofessional framework that recognizes the interconnectedness of human, animal, and environmental health and promotes coordinated actions to optimize health outcomes across all domains (CDC 2023; WHO 2022). This approach is supported by evidence that over 60% of human infectious diseases are zoonotic, with their emergence and transmission strongly influenced by environmental factors (Prata 2022; O'Brien and Xagoraraki 2019). The concept originated from "One Medicine," introduced by Calvin Schwabe, which emphasized collaboration between veterinary and human medicine for disease control and surveillance (Schwabe 1986).

The modern One Health paradigm extends this framework to explicitly include environmental health, highlighting shared risks among humans, animals, and ecosystems. Water, as a common resource, plays a central role in disease transmission and chemical exposure. Contaminated water can harbor pathogenic microorganisms such as *Escherichia coli*, *Salmonella* spp., and *Vibrio cholerae*, affecting human populations, livestock, and wildlife (O'Brien and Xagoraraki 2019; Prata 2022; Kumar et al. 2023). Additionally, water acts as a transport medium for chemical contaminants, including pesticides, pharmaceuticals, heavy metals, and microplastics, which bioaccumulate and impact food webs across species (Falkenmark et al. 2019; Prata 2022).

The One Health approach to water is founded on four core principles: interdisciplinarity, integrating medical, veterinary, ecological, and environmental expertise (WHO 2022); surveillance and monitoring of pathogens, chemical contaminants, and antimicrobial resistance (AMR) across interconnected reservoirs (Prata 2022; O'Brien and Xagoraraki 2019); control and treatment strategies to mitigate risks to humans, animals, and ecosystems (Pitt and Gunn 2024); and sustainability, ensuring long-term ecosystem integrity alongside human and animal health (Falkenmark et al. 2019). Applied to water systems, One Health enables integrated risk assessment and management, particularly in regions facing water scarcity, inadequate sanitation, and intensive agricultural pressures (Prata 2022; Pitt and Gunn 2024).

6.1.2 Global Water Challenges in the Anthropocene

The Anthropocene represents a proposed geological epoch characterized by pervasive human influence on Earth's natural systems, including freshwater resources (Rockström et al. 2009; Vörösmarty et al. 2010). Rapid urbanization, industrialization, and agricultural intensification have profoundly altered hydrological cycles, affecting both the availability and quality of freshwater worldwide (Prüss-Ustün et al. 2019; Falkenmark et al. 2019). Water pollution has consequently emerged as a critical global concern, with industrial effluents, untreated sewage, and agricultural runoff introducing nutrients, heavy metals, pathogens, and emerging contaminants into surface water and groundwater systems (Prata 2022).

Nutrient enrichment from fertilizers promotes eutrophication and algal blooms, leading to hypoxia and biodiversity loss in aquatic ecosystems (Falkenmark et al. 2019). In parallel, emerging contaminants such as pharmaceuticals and microplastics pose chronic risks to human and animal health and disrupt ecosystem functioning (Franklin et al. 2024; *The Guardian* 2024). Water scarcity further amplifies these challenges. Nearly 90% of the global population experiences some degree of water stress, with the most severe impacts observed in South Asia, Sub-Saharan Africa, and Latin America (Falkenmark et al. 2019; Prüss-Ustün et al. 2019). The combined pressures of water scarcity, pollution, and population growth intensify health inequities, threaten food security, and impose substantial economic burdens (Rockström et al. 2009; Vörösmarty et al. 2010).

Against this backdrop, this chapter examines the interlinkages between water systems and health through the One Health framework, emphasizing integrated strategies to protect human, animal, and environmental health. Specifically, it focuses on: (1) water quality within a One Health context; (2) drivers of water quality deterioration; (3) waterborne diseases and antimicrobial resistance (AMR); and (4) approaches for water quality monitoring and assessment. The need for interdisciplinary collaboration and comprehensive management strategies is underscored by the central role of water quality in shaping health outcomes in the Anthropocene, as conceptualized within the One Health paradigm (Musicha et al. 2024; Falkenmark et al. 2019). Conceptual schematic of the One Health approach: integrating environmental, animal, and human health across scales (local to global) via collaboration, coordination, and communication, enabling resilient and equitable outcomes, is illustrated in Fig. 6.1.

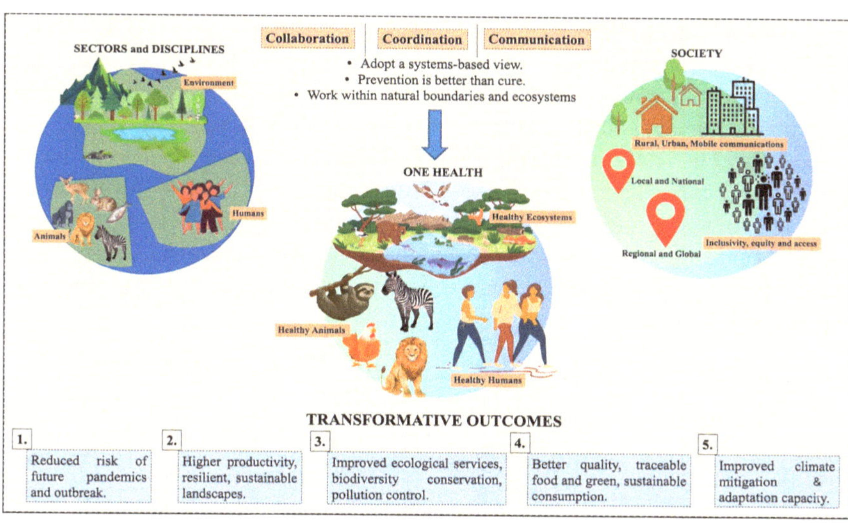

Fig. 6.1 Conceptual diagram showing the *One Health* framework linking human, animal, and environmental Health

6.2 Understanding Water Quality in a *One Health* Context

6.2.1 Defining Water Quality: Chemical, Microbial, and Ecological Parameters

Water bodies are complex systems governed by interacting physical, chemical, biological, and ecological processes, enabling their use for drinking water supply, agriculture, recreation, and biodiversity conservation (Gleick 2014). Within the One Health framework, water quality is evaluated by jointly considering human health, animal well-being, and ecosystem integrity. Chemical stressors include nutrients (nitrogen and phosphorus), heavy metals (e.g., lead, mercury, and arsenic), and organic pollutants (e.g., pesticides and pharmaceuticals). Excessive nutrient loading drives eutrophication and hypoxia, resulting in fish mortality and ecosystem degradation, with cascading effects on food security for human and livestock populations (Díaz and Rosenberg 2008). Heavy metals and persistent organic pollutants bioaccumulate in aquatic organisms and biomagnify through food webs, posing long-term health risks to humans and wildlife (Järup 2003; Jaishankar et al. 2014). Table 6.1 synthesizes priority water-quality parameters used in One Health assessments—spanning microbial indicators, nutrients, cyanotoxins, disinfection by-products, metals, pesticides, and emerging contaminants—linking typical sources, cross-sector relevance (human–animal–environment), common analytical approaches, and guideline values.

Microbial contamination is a critical indicator of waterborne health risk. Pathogens such as *Escherichia coli*, *Salmonella* spp., *Vibrio cholerae*, and *Cryptosporidium* circulate among humans, livestock, and wildlife, facilitating zoonotic disease transmission. Global water quality surveillance demonstrates persistent microbial contamination, underscoring challenges in achieving Sustainable Development Goal 6 (SDG 6) for safe water access (Rose et al. 2023). In addition, the detection of antimicrobial resistance (AMR) genes in water bodies highlights the role of aquatic environments as reservoirs and transmission pathways for resistance across human, animal, and environmental interfaces (Prata 2022).

Ecological indicators reflect the biological integrity of water bodies and include species diversity, abundance, and the presence of bioindicator organisms such as macroinvertebrates. Declining biodiversity signals the cumulative impacts of anthropogenic pressures, including pollution, habitat alteration, and climate change, which indirectly compromise human and animal health by disrupting ecosystems (Vörösmarty et al. 2010).

Table 6.1 Key water quality parameters and their relevance to One Health

Parameter/example analytes	Typical sources	One Health relevance (human, animal, environment)	Common methods/ notes	Typical guideline
Microbial indicators (WHO 2024; **WHO; Leonard and Eaton** 2021; **of Science** 2009a, b)				
Escherichia coli (E. coli)	Human/animal feces; sewage leaks; runoff from manure; septic failures	Signals fecal contamination and potential enteric pathogens affecting people, livestock, and wildlife; contamination of irrigation water links to foodborne outbreaks	Membrane filtration culture (EPA Method 1603; ISO 9308-1); presence/absence and IDEXX methods also used	Drinking-water: E. coli not detectable in 100 mL (WHO)
Enterococci	Human/animal feces; sewage-contaminated recreational waters	Reliable indicator of health risk in marine & fresh recreational waters; tracks zoonotic contamination	Culture by membrane filtration (EPA Method 1600); qPCR methods used in some programs	Recreational waters (US EPA, 2012 RWQC): GM 35 CFU/100 mL (marine)/30 CFU/100 mL (freshwater); STV 130 CFU/100 mL
Nutrients (Nitrate and Nitrite in Drinking Water, 2025)				
Nitrate (NO_3^-)	Fertilizer runoff; animal feeding operations; septic systems	Methemoglobinemia risk for infants; linked to eutrophication harmful algal blooms that affect humans, livestock, wildlife	Colorimetry (cadmium reduction), ion chromatography; manage catchment sources	50 mg/L as nitrate (≈11 mg/L as nitrate-N) in drinking-water (WHO)
Nitrite (NO_2^-)	Nitrification in distribution systems; wastewater inputs; chloramination	Methemoglobinemia can indicate nitrification episodes that also favor microbial regrowth	Colorimetry: manage disinfectant residuals to prevent nitrification	3 mg/L as nitrite (≈0.9 mg/L as nitrite-N) in drinking water (WHO)
Cyanobacterial toxins (*Directive - 2020/2184 - EN - EUR-Lex*)				

(continued)

Table 6.1 (continued)

Parameter/example analytes	Typical sources	One Health relevance (human, animal, environment)	Common methods/ notes	Typical guideline
Microcystin-LR (representative microcystins)	Eutrophication-driven cyanobacterial blooms in lakes/ reservoirs; warm, nutrient-rich waters	Hepatotoxins cause illness in people, livestock & pets, resulting in deaths from bloom ingestion; ecosystem impacts	ELISA screening; LC-MS/MS confirmation; bloom surveillance via remote sensing + field sampling	1 μg/L microcystins in drinking-water (WHO; also EU DWD)
Disinfection byproducts (Stage 1 and Stage 2 Disinfectants and Disinfectants US EPA)				
Total trihalomethanes (TTHMs: chloroform, bromodichloromethane, dibromochloromethane, bromoform)	Reaction of chlorine/ ozone with natural organic matter & bromide	Chronic exposure linked to cancer & reproductive risks; balance microbial risk vs. DBP risk	Purge-and-trap GC/ MS; manage precursors and optimize disinfection	80 μg/L (EPA Stage 2 DBPR, running annual average); EU parametric value 100 μg/L
Haloacetic acids (HAA5)	Chlorination of waters rich in natural organic matter	Chronic exposure concerns (carcinogenicity, developmental effects)	Derivatization + GC/ MS; similar control strategies as TTHMs	60 μg/L (EPA Stage 2 DBPR, running annual average)
Heavy metals/metalloids (Arsenic 2025; **OMS** 2022)				
Arsenic (As)	Geogenic release (arsenic-bearing aquifers); mining; geothermal; industrial discharges	Carcinogen; cardiovascular, skin, neurodevelopmental effects; accumulates in food chain; wildlife toxicity	ICP-MS; adsorptive media (Fe/Al); coagulation/filtration; source substitution	10 μg/L (WHO provisional drinking-water guideline)
Lead (Pb)	Corroding lead service lines, premise plumbing, solder, and fittings	Neurotoxin (no safe exposure level) with developmental harms; domestic animals are also affected	First-draw sampling; corrosion control; lead service line replacement	EU DWD: 5 μg/L parametric value (to apply by 2036); WHO emphasizes lowering exposure as far as possible

Parameter/example analytes	Typical sources	One Health relevance (human, animal, environment)	Common methods/ notes	Typical guideline
Mercury (Hg)	Industrial discharges; artisanal mining; atmospheric deposition → methylation in aquatic systems	Neurotoxin; methylmercury bioaccumulates in fish, exposing humans/wildlife via diet	Cold vapor atomic fluorescence/ICP-MS; source reduction	EU DWD: 1 µg/L parametric value (drinking-water)
Cadmium (Cd)	Batteries, pigments; pipe/solder leaching; phosphate fertilizers	Kidney toxicity; bone effects; accumulates in crops and aquatic organisms	ICP-MS; corrosion/ source control	WHO drinking-water guideline: 3 µg/L
Pesticides (General & Aquatic life) (*Directive - 2020/2184 - EN - EUR-Lex; Aquatic Life Benchmarks. US EPA*)				
Any single pesticide; Total pesticides	Agricultural runoff; improper storage/ handling; vector control	Endocrine, neuro- and developmental toxicity; pollinator and aquatic invertebrate harm; food chain impacts	Targeted LC-MS/MS suites; passive samplers; catchment BMPs	EU DWD: 0.1 µg/L (any single pesticide), 0.5 µg/L (sum of pesticides)
Imidacloprid (neonicotinoid)	Seed treatments; foliar & soil applications → runoff/tiles/ditches	Aquatic invertebrate toxicity → food web effects; potential indirect impacts on fisheries & biodiversity	LC-MS/MS; interpret vs. aquatic life benchmarks	US EPA Aquatic Life Benchmarks (freshwater invertebrates): chronic ≈0.01 µg/L; acute ≈0.385 µg/L (illustrative)
Emerging contaminants (PFAS) (pharmaceuticals & endocrine disruptors) (AMR markers) (microplastics)				
PFOA, PFOS, PFNA, PFHxS, HFPO-DA (GenX), PFBS (for hazard index) (11) (US EPA, Pre-Publication Version; *PFAS National Primary Drinking Water Regulation; Per- and Polyfluoroalkyl Substances (PFAS)*)	AFFF foams; industrial discharges; textiles; food packaging; biosolids	Persistent, bioaccumulative, immunotoxicity and other chronic effects; detected in wildlife; crosses media (water–soil–biota)	LC-MS/MS (EPA 533/537.1); precursor oxidation tests; granular activated carbon/ion exchange/ RO for control	US EPA (Final NPDWR, 2024): PFOA 4 ng/L; PFOS 4 ng/L; PFNA 10 ng/L; PFHxS 10 ng/L; HFPO-DA (GenX) 10 ng/L; mixture Hazard Index of 1. EU DWD: sum of 20 PFAS 0.1 µg/L; PFAS total 0.5 µg/L

(continued)

Table 6.1 (continued)

Parameter/example analytes	Typical sources	One Health relevance (human, animal, environment)	Common methods/ notes	Typical guideline
Diclofenac; 17-β-estradiol (E2); 17-α-ethinylestradiol (EE2) (Huang et al. 2015)	Wastewater effluent; hospital/household disposal; livestock operations	Fish feminization; renal/ hepatic effects in aquatic organisms; potential chronic human effects at low levels are still under study	LC-MS/MS at ng/L; effect-based bioassays (ER-Calux) as screening	EU Water Framework Directive Watch List (for environmental monitoring); no WHO drinking-water guideline values
intI1 (class 1 integron), sul1, blaCTX-M, mcr-1, tetA (genes); ESBL-*E. coli* (organism)	Wastewater, hospitals, livestock manure, aquaculture, and combined sewer overflows	Water is a pathway for resistant bacteria/genes between humans, animals, and environments; surveillance informs cross-sector interventions	qPCR/ddPCR for marker genes; culture for ESBL-*E. coli* (WHO Tricycle); metagenomics for resistome profiling	No drinking-water guideline; risk increases when antibiotic concentrations exceed PNEC-resistance thresholds (e.g., ciprofloxacin ≈0.064 μg/L)
Particles >1 μm	Fragmented plastics; textiles; tire wear; wastewater effluent	Ingestion by aquatic life; vector for additives/adsorbed pollutants; human health risks still being assessed	Spectroscopy (μ-FTIR/ Raman); clean-lab protocols to avoid contamination	No WHO drinking-water guideline value; WHO recommends risk-based management while evidence evolves

6.2.2 Links Between Water, Ecosystems, and Public Health

Since disease regulation, drinking water, fisheries, and irrigation are the most essential facilities provided by the Health water system, contaminated water impacts these facilities. It causes direct or indirect impacts on well-being (Gleick 2014). For instance, direct impacts of compromised water quality, evidenced by gastrointestinal infections, hepatitis, and cholera outbreaks. Conversely, indirect impacts include pollution effects on aquatic biodiversity, reduced fisheries productivity, and the ability of invasive species to thrive, thereby altering ecosystem dynamics. Vulnerable populations are often disproportionately affected by compromised water quality, especially in low- and middle-income countries where sanitation infrastructure is limited (Rose et al. 2023; WHO 2017). One health perspective emphasizes the interlinked relationships among environmental degradation, animal health, and human morbidity and mortality. As shown in Fig. 6.2, pollutants from climate extremes, urban–industrial activities, agriculture, and wastewater enter groundwater, surface, coastal, and engineered water systems, undergo transport and transformation (e.g., sorption, biodegradation, volatilization, bioaccumulation, gene transfer), and ultimately affect animal, ecosystem, and human health, with feedbacks such as wetland loss and antimicrobial resistance.

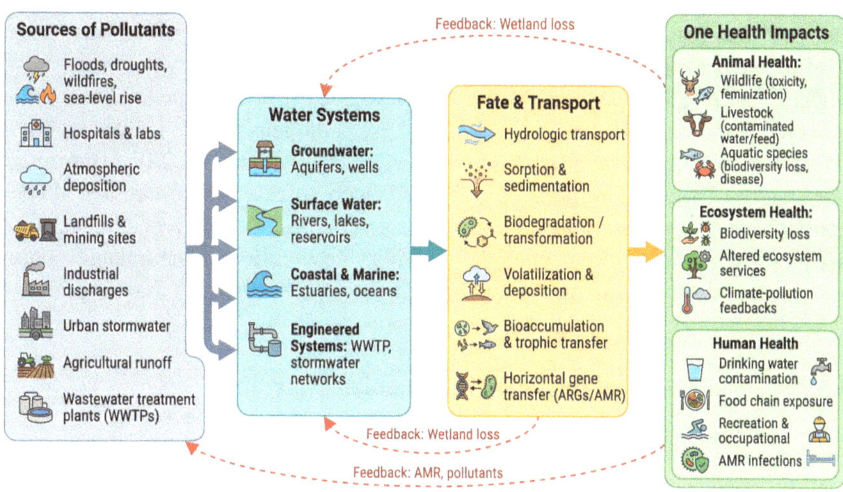

Fig. 6.2 Pathways of pollutants entering water systems and their One Health impacts

6.2.3 *Case Studies of Waterborne Diseases Impacting Humans, Animals, and Agriculture*

6.2.3.1 Cholera Outbreaks in South Asia

Vibrio cholerae is the causative agent of Cholera, a disease of enduring concern across multiple regions of South Asia. These bacteria propagate predominantly in contaminated waters, particularly in areas with inadequate sanitation systems. Acceleration in its transmission by disseminating bacteria through rivers and transient flows is intensified by flooding and monsoon cycles (Sugimoto et al. 2014). Subsequently, human food chains are also affected when contaminated animal-source water invades, as livestock consuming this untreated water increases the potency of zoonotic or agricultural contamination. Control measures mainly depend on environmental surveillance, such as temperature and hydrological conditions, to anticipate and prevent outbreaks, and outline the need for a collaborative One Health preparedness workflow.

6.2.3.2 Cryptosporidiosis Associated with Livestock

Among waterborne diseases, Cryptosporidiosis is among the most threatening, caused by the most resilient pathogen, *Cryptosporidium,* a protozoan parasite (Caffarena et al. 2020). A study in Kenya linked cattle rearing with compromised surface water used by society. Shared water sources are contaminated by livestock movement and fecal discharge. Oocysts of these pathogens remain viable and highly infectious in water for approximately half a year (Floyd et al. 2022). According to one study in Uruguay, millions of individuals were affected by contaminated water containing zoonotic Cryptosporidium parvum subtypes released by cattle feces into drainage streams (Caffarena et al. 2020). This study reinforces the linkages among livestock, hydrology, and human well-being, as well as waterway management.

6.2.3.3 Pesticide Contamination in Ethiopian Watersheds

A study of Ethiopian highlands reported widespread pesticide contamination in both surface and groundwater within smallholder farming groups. Various common pesticides, such as chlorpyrifos, endosulfan, and dimethoate, were detected in irrigation and household water (Sishu et al. 2022). Due to a lack of knowledge about the Health and environmental risks associated with pesticides, humans are exposed to these contaminants and face serious health threats. These contaminants affect multiple dimensions, including aquatic biodiversity, human neurological health, and agricultural livelihoods when contaminated water is repeatedly used. The study emphasizes the necessity of collaborative strategies, such as One Health,

Table 6.2 Examples of pathogens and resistance genes detected in water and their One Health implications

Pathogen	Resistance genes detected	Human/animal impact
E. coli	blaCTX-M, tetA	Gastroenteritis, urinary infections
Klebsiella spp.	blaNDM, blaKPC	Sepsis, pneumonia
Salmonella spp.	blaTEM, qnrS	Typhoid, gastroenteritis
Vibrio cholerae	ctxAB	Cholera
Shigella spp.	sul1, dfrA	Dysentery

comprising societal education, control of agrochemical use, and integrated watershed management.

6.2.3.4 Cholera in Haiti: Post-disaster Waterborne Outbreak

In 2010, Haiti experienced a severe outbreak of Cholera. The Artibonite River, a crucial water source for day-to-day tasks, was the cause of the epidemic. A UN peacekeeper contingent connected to human waste releases into the river, and an inadequate sanitation system compromised water quality by contaminating it. The presence of *Vibrio cholerae* in co-utilized water resources can rapidly affect humans. Post-crisis contexts can be addressed through a single health system, as this study indicates the impacts of environmental disturbance, social dynamics, and cross-domain water use (Piarroux et al. 2011).

6.2.3.5 Cyanobacterial Toxins and Wildlife Mortality

This is the case of morbidity in more than 350 elephants after consuming water from waterholes contaminated with blue-green algae (cyanobacteria), a consequence of climate-induced eutrophication and heat stress. Rapid cyanobacterial growth was evidenced in satellite imagery of water bodies due to drought and subsequent heavy rains. This case clearly illustrates the impact of climate stressors on water quality and the acute threats they pose to biodiversity and ecosystems (Lomeo et al. 2024). Table 6.2 lists representative waterborne pathogens and clinically relevant resistance genes reported in environmental surveillance, highlighting the potential human and animal health outcomes associated with exposure.

6.3 Drivers of Water Quality Deterioration

Compromised water quality results from a range of anthropogenic and natural factors. These factors contribute to waterborne diseases, along with poor ecosystem conditions. These waterborne diseases, due to poor water quality, often cause antimicrobial resistance (AMR), bridging the gap between human, animal, and environmental Health.

6.3.1 *Urbanization, Industrialization, and Agricultural Intensification*

Urbanization creates resistant surfaces that further inhibit natural infiltration and increase the efflux of pollutants into rivers, groundwater, and lakes (Paul and Meyer 2001). However, pollutants from industrialization, including heavy metals, thermal products, and chemical waste, when they exceed permissible limits, cause uncommon consequences (Järup 2003). Excessive use of herbicides and pesticides introduces high levels of nitrogen and phosphorus into waterways through intensive agriculture, causing eutrophication and, in turn, harmful algal blooms due to dissolved oxygen depletion. Aquatic life and water quality deteriorated as the aforementioned conditions (Conley et al. 2009). Livestock fecal efflux can also carry pathogen and antibiotic residues, which can be introduced into nearby water bodies and facilitate AMR propagation (Zhu et al. 2013). For instance, China's Lake Tai Basin is severely affected by intensive agricultural practices, leading to repeated cyanobacterial blooms driven by excess nutrients and resulting in poor water quality and reduced aquatic biodiversity. Conversely, groundwater in India and South Asia has been found contaminated by industrial effluents (Paerl et al. 2011).

6.3.2 *Climate Change Impacts on Hydrology and Contaminant Transport*

Climate change has a significant impact on precipitation patterns, sometimes leading to extreme weather, and on river flow, groundwater recharge, water quality, and pollutant transport (IPCC 2022). Pollutants from multiple sources, such as soils, sediments, and urban areas, can enter water bodies due to heavy rainfall and flooding. However, in shrinking water bodies, contamination results from droughts (IPCC 2022; Dogra et al. 2024). For example, in Bangladesh in 2020, extensive floods led to the introduction of pollutants such as sewage, livestock feces, and chemical waste into rivers, causing waterborne outbreaks of Cholera and dysentery. According to a report by the International Federation of Red Cross and Red Crescent Societies, flooding in Bangladesh affected around 4.5 million people in the city. The significant consequences of this flood were waterborne diseases such as Cholera and dysentery (IFRC 2020). Moreover, previous research has reported cases of enterotoxigenic Escherichia coli (ETEC) infection during flooding. However, ETEC and Cholera are considered major contributors to diarrhoea during floods, reinforcing the necessity of proper management of both water quality and human health treatments (Rupa et al. 2024).

Infectious pathogens, antibiotic residues, pharmaceutical waste, and microplastics generally remain in untreated or partially treated wastewater. These wastes make bacteria resistant to survive in such a toxic environment (Michael et al. 2013). Microplastics adsorb chemical pollutants and contain microbes, facilitating the transmission of pathogens across ecosystems (Su et al. 2021; Yang et al. 2019). For instance, in South Korea, wastewater treatment plants exhibited multi-drug resistance due to high concentrations of antibiotics such as sulfonamides and tetracyclines, resulting from multidrug-resistant *E. coli* in rivers (Park et al. 2024).

6.4 Waterborne Diseases and Antimicrobial Resistance (AMR)

6.4.1 Historical and Contemporary Outbreaks Linked to Poor Water Quality

Since the nineteenth century, multiple pandemics of Cholera have occurred through the transmission of the Cholera-causing agent *Vibrio cholerae* into compromised water. These bacteria are primarily transmitted through contaminated water, as this concept was initially introduced by John Snow's work during the 1854 London outbreak. WHO reported more than 10,000 individual deaths in the 2010 outbreak of Haiti, a consequence of infrastructure collapse due to the earthquake (Piarroux et al. 2011; Ocasio et al. 2023). A recent outbreak occurred in 2024 in Sudanese refugee camps due to limited access to clean water and overcrowding of water sources (The Guardian 2024). *Salmonella enterica* serotype Typhi causes typhoid fever. Inadequate sanitation and the additional burden of overcrowding in urban areas led to the development of this condition. There is a well-known case of an asymptomatic carrier, known as "Typhoid Mary". Pakistan has previously experienced typhoid in 2018, which was extensively drug-resistant and made intervention complicated (Pakistan Health Authorities 2018). A recent outbreak emerged in 2024 in the UK, linked to travel from South Asia, with a total of 702 reported cases (UKHSA 2025). This data reinforced the urgent need for vaccine administration, better sanitation, and education about this. Also, *Shigella* species or *Entamoeba histolytica* are the pathogens responsible for dysentery, which primarily causes morbidity in overcrowded areas. Jaipur experienced an outbreak in 2023, many cases were reported to be hospitalized, and at least one death occurred due to inappropriate management of wells and contaminated water.

6.4.2 *Role of Wastewater and Environmental Reservoirs in AMR and Pathogen Dissemination*

Antimicrobial-resistant bacteria and genes persist in wastewater (Sambaza and Naicker 2023). Human Health is the primary concern due to contaminated water consumption. Antibiotics and other selective agents in wastewater support the proliferation and transmission of such microbes, which can invade natural water bodies and pose health risks to both humans and animals (Fouz 2020). AMR transmission through wastewater plants is evidenced by environmental surveillance and the presence of many multidrug-resistant *E. coli* and *Klebsiella* spp. in rivers receiving untreated effluents (Meradji 2025). Horizontal gene transfer facilitates the transfer of ARGs across multiple pathogens. Various reservoirs, such as soil, water bodies, and sediments, harbor microbial communities that comprise antibiotic- and antibiotic-resistant bacteria (Zhu et al. 2025; Popoola 2025). Environmental reservoirs are the primary drivers of the persistence and spread of AMR across ecological and human systems. Figure 6.3 conceptualizes antimicrobial resistance (AMR) transmission in a One Health framework, showing how antibiotic use in humans and animals generates ARB/ARGs that circulate through environmental reservoirs and water systems, with exposure pathways back to humans via drinking water, food chains, and recreation.

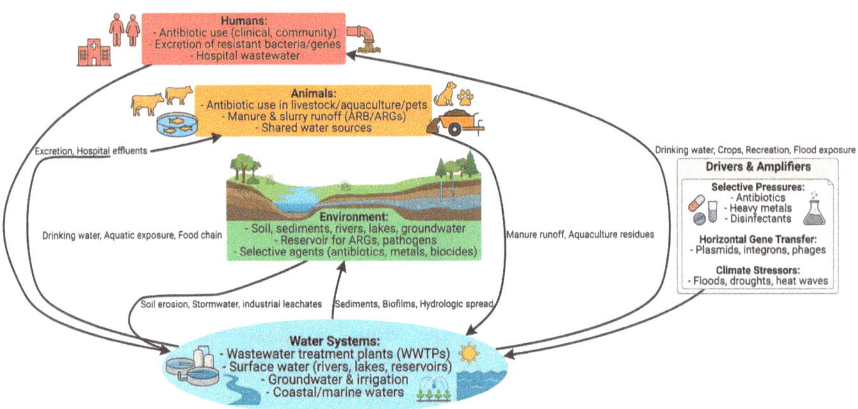

Fig. 6.3 Cycle of AMR transmission through water across human, animal, and environmental interfaces

6.5 Monitoring and Assessment of Water Quality

6.5.1 Traditional vs. Advanced Monitoring Techniques

Traditional water quality monitoring has relied on culture-based techniques, such as plate counts and membrane filtration, to enumerate indicator bacteria. However, these methods are constrained by long incubation periods (typically 24–48 h), inability to detect non-culturable microorganisms, and reliance on statistical extrapolation from limited sample sizes (Suttner 2020; Ong et al. 2020). To overcome these limitations, molecular approaches—including PCR, quantitative PCR (qPCR), and fluorescence in situ hybridization (FISH)—enable rapid and sensitive detection of specific pathogens or indicator genes. Despite their advantages, these targeted assays provide limited insight into overall microbial community structure.

Metagenomic approaches, including shotgun sequencing and 16S rRNA amplicon profiling, offer a comprehensive, culture-independent characterization of microbial communities, revealing both known and previously unrecognized taxa as well as functional gene potential in water systems (Rout et al. 2025). Metagenomics has demonstrated particular value in reclaimed water surveillance, enabling non-targeted monitoring of biological contaminants when combined with complementary tools such as flow cytometry and advanced bioinformatics pipelines (Hong et al. 2020). Studies of drinking water systems consistently show that culture-based methods substantially underestimate microbial diversity, whereas high-throughput sequencing provides more accurate insights into microbiome structure and contamination dynamics (Brumfield et al. 2020).

Integrating conventional cultivation with molecular and metagenomic tools—an approach referred to as *culturomics*—further enhances microbial detection. Culturomics employs high-throughput cultivation and mass spectrometry techniques, such as MALDI-TOF, to characterize previously uncultivated organisms, thereby improving the completeness and interpretability of metagenomic datasets. Emerging multi-omics frameworks that integrate genomics, transcriptomics, and metabolomics are poised to link microbial identity to functional activity, advancing predictive, risk-based monitoring of water systems. Collectively, these advanced approaches provide greater sensitivity, broader coverage, and more actionable information than traditional monitoring, strengthening the management of microbial risks to water quality.

6.5.2 *Wastewater-Based Epidemiology (WBE) for Surveillance*

Wastewater-based epidemiology (WBE) quantifies chemical and biological signals in pooled sewage samples collected from sewer networks or wastewater treatment plants (WWTPs) to infer population-level public health indicators. Measured concentrations of target compounds provide insights into community-wide substance use, chemical exposures, disease prevalence, and antimicrobial resistance (AMR). One of the earliest proposed applications of WBE was the analysis of wastewater to estimate illicit drug consumption (Daughton 2001). Since then, WBE has expanded to include a wide range of indicators, such as alcohol and tobacco use, pharmaceuticals, pathogens, therapeutic agents, and AMR markers (Boogaerts et al. 2021; Senta et al. 2020; Sims and Kasprzyk-Hordern 2020).

To date, WBE has primarily served as a complementary surveillance tool rather than a primary driver of public health interventions. Nevertheless, its broad applicability underscores its potential to support a One Health approach that recognizes the interconnectedness of human, animal, and environmental health. Embedding WBE within integrated public health surveillance systems will require leadership from public health institutions and strategic program design. The use of multicriteria decision-making frameworks has been proposed as a promising approach to ensure that WBE programs address current health priorities while remaining adaptable to emerging indicators (Gunnar and Kankaanpää 2019; Keshaviah et al. 2021).

A significant challenge lies in integrating heterogeneous data streams to improve interpretation, identify cross-sectoral linkages, and inform actionable responses. Achieving this integration demands coordinated collaboration among public health authorities, researchers, policymakers, clinicians, environmental scientists, wastewater utilities, and non-traditional stakeholders such as law enforcement. Such a multi-stakeholder governance model aligns with One Health principles by explicitly accounting for interactions between environmental exposures, human health outcomes, and societal systems. Effective implementation requires clearly defined objectives, transparent communication pathways, and a central coordinating body for data management and interpretation (Lancaster and Rhodes 2020). Further research is needed to assess cost-effectiveness, define contexts in which WBE offers greater resolution or efficiency than conventional surveillance, and quantify its added value across diverse One Health applications. Generating this evidence will be critical for transitioning WBE from a supportive monitoring tool to a proactive component of integrated health surveillance and response systems.

6.6 Conclusion

Water represents a critical interface where the interconnected health of humans, animals, and ecosystems becomes most evident, and the One Health framework provides an integrated lens to address these linkages. This chapter has demonstrated how chemical contamination, microbial pollution, and ecological degradation interact within water systems to drive disease emergence, antimicrobial resistance (AMR), and biodiversity loss. Case studies spanning cholera outbreaks in South Asia and Haiti, pesticide poisoning in Ethiopia, and cyanobacterial blooms affecting wildlife illustrate the complex and escalating nature of waterborne health risks under intensifying anthropogenic pressures. Urbanization, agricultural intensification, industrial discharges, climate change, and untreated wastewater collectively contribute to water quality deterioration, with impacts extending across species and ecosystems.

The persistence of AMR in environmental reservoirs and the rising prevalence of emerging contaminants highlight the limitations of discipline-specific approaches to water management. Addressing these challenges requires integrated solutions that treat human, animal, and environmental health as inseparable components of a shared system. Emerging surveillance tools—including molecular diagnostics, metagenomics, and wastewater-based epidemiology (WBE)—offer unprecedented capacity to detect pathogens, pollutants, and resistance genes with enhanced speed and sensitivity. However, their practical application depends on interdisciplinary collaboration, equitable access to technology, and the translation of scientific data into actionable policy.

In the Anthropocene, where water scarcity, pollution, and population growth converge to amplify global health risks, safeguarding water quality has become central to health security. The One Health paradigm underscores the necessity of collective, preventive, and adaptive strategies that integrate scientific innovation with governance, education, and community engagement. Through this approach, societies can advance toward resilient water governance systems that protect public health, secure food systems, preserve biodiversity, and enhance preparedness for future environmental and health challenges.

References

Aquatic Life Benchmarks and Ecological Risk Assessments for Registered Pesticides. US EPA. Available from: https://www.epa.gov/pesticide-science-and-assessing-pesticide-risks/aquatic-life-benchmarks-and-ecological-risk? Accessed 16 Sept 2025

Arsenic. Available from: https://www.who.int/news-room/fact-sheets/detail/arsenic?utm. Accessed 16 Sept 2025

Boogaerts T, Ahmed F, Choi PM, Tscharke B, O'Brien J, De Loof H et al (2021) Current and future perspectives for wastewater-based epidemiology as a monitoring tool for pharmaceutical use. Sci Total Environ 789:148047

Brumfield KD, Hasan NA, Leddy MB, Cotruvo JA, Rashed SM, Colwell RR, Huq A (2020) A comparative analysis of drinking water employing metagenomics. PLoS One 15(4):e0231210

Caffarena RD, Meireles MV, Carrasco-Letelier L, Picasso-Risso C, Santana BN, Riet-Correa F, Giannitti F (2020) Dairy calves in Uruguay are reservoirs of zoonotic subtypes of *Cryptosporidium parvum* and pose a potential risk of surface water contamination. Front Vet Sci 7:562. https://doi.org/10.3389/fvets.2020.00562

CDC (2023) One Health basics. Centers for Disease Control and Prevention. https://www.cdc.gov/one-health/about/index.html

Conley DJ, Paerl HW, Howarth RW, Boesch DF, Seitzinger SP, Havens KE, Lancelot C, Likens GE (2009) Ecology. Controlling eutrophication: nitrogen and phosphorus. Science, New York, N.Y. 323(5917):1014–1015. https://doi.org/10.1126/science.1167755

Daughton CG (2001) Illicit drugs in municipal sewage. Pharm Care Products Environ 791:348–364

Díaz RJ, Rosenberg R (2008) Spreading dead zones and consequences for marine ecosystems. Science 321(5891):926–929. https://doi.org/10.1126/science.1156401

Dogra K, Lalwani D, Dogra S, Panday DP, Raval N, Trivedi M et al (2024) Indian and global scenarios of bisphenols a distribution and its new analogues: prevalence & probability exceedance. J Hazard Mater 135128. https://doi.org/10.1016/j.jhazmat.2024.135128

Falkenmark M, Wang-Erlandsson L, Rockström J (2019) Understanding of water resilience in the Anthropocene. J Hydrol X 2:100009

Floyd JR, Kwoba E, Mwangi T, Okotto-Okotto J, Wanza P, Wardrop N, Yu W, Wright JA (2022) A spatiotemporal analysis of cattle herd movement in relation to drinking-water sources: implications for Cryptosporidium control in rural Kenya. Environ Sci Pollut Res 29(23):34314–34324. https://doi.org/10.1007/s11356-021-17888-3

Fouz N (2020) The contribution of wastewater to the transmission of AMR. PMC Rev

Franklin AM, Weller DL, Durso LM, Bagley M, Davis BC, Frye JG et al (2024) A One Health approach for monitoring antimicrobial resistance: developing a national freshwater pilot effort. Front Water 6:1359109. https://doi.org/10.3389/frwa.2024.1359109

Gleick PH (ed) (2014) The world's water volume 8: the biennial report on freshwater resources. Island Press

Gunnar T, Kankaanpää A (2019) The practical implications of wastewater-based illicit drug epidemiology. Curr Opin Environ Sci Health 9:49–57

Hong PY, Mantilla-Calderon D, Wang C (2020) Metagenomics as a tool to monitor reclaimed-water quality. Appl Environ Microbiol 86(16):e00724–e00720

Huang B, Sun W, Li X, Liu J, Li Q, Wang R, Pan X (2015) Effects and bioaccumulation of 17β-estradiol and 17α-ethynylestradiol following long-term exposure in crucian carp. Ecotoxicol Environ Saf 112:169–176. https://doi.org/10.1016/j.ecoenv.2014.11.004

IFRC (2020) 17.5 million affected by floods and threatened by disease in South Asia. International Federation of Red Cross and Red Crescent Societies. https://www.ifrc.org/press-release/175-million-affected-floods-and-threatened-disease-south-asia

IPCC (2022) Climate change 2022: impacts, adaptation, and vulnerability. Intergovernmental Panel on Climate Change. https://www.ipcc.ch/report/ar6/wg2/

Jaishankar M, Tseten T, Anbalagan N, Mathew BB, Beeregowda KN (2014) Toxicity, mechanism and health effects of some heavy metals. Interdiscip Toxicol 7(2):60–72. https://doi.org/10.2478/intox-2014-0009

Järup L (2003) Hazards of heavy metal contamination. Br Med Bull 68(1):167–182. https://doi.org/10.1093/bmb/ldg032

Keshaviah A, Hu XC, Henry M (2021) Developing a flexible national wastewater surveillance system for COVID-19 and beyond. Environ Health Perspect 129(4):045002

Kumar M, Mazumder P, Silori R, Manna S, Panday DP, Das N et al (2023) Prevalence of PPCPs, microplastics and co-infecting microbes in the post-COVID-19 era and its implications on antimicrobial resistance and potential endocrine disruptive effects. Sci Total Environ:166419–166419

Lancaster K, Rhodes T (2020) Wastewater monitoring of SARS-CoV-2: lessons from illicit drug policy. Lancet Gastroenterol Hepatol 5(7):641–642

Leonard M, Eaton C (2021) Recreational water quality guidelines update. Environmental Science and Research Institute Client Report FWFW21029 for Ministry of Health. https://www.esr.cri.nz/media/seap4u1y/recreational-water-quality-guidelines-update.pdf

Lomeo D, Tebbs E, Babayani ND, Chadwick M, Gondwe M, Jungblut AD, McCulloch GP, Morgan ER, Schillereff D, Simis SGH, Songhurst AC (2024) Remote sensing and spatial analysis reveal unprecedented cyanobacteria bloom dynamics associated with elephant mass mortality. Sci Total Environ 957:177525. https://doi.org/10.1016/j.scitotenv.2024.177525

Meradji S (2025) The role of water as a reservoir for antibiotic-resistant genes. PMC

Michael I, Rizzo L, McArdell CS et al (2013) Urban wastewater treatment plants as hotspots for the release of antibiotics in the environment: a review. Water Res 47(3):957–995. https://doi.org/10.1016/j.watres.2012.11.027

Musicha P et al (2024) Time to define One Health approaches to tackling antimicrobial resistance. Nat Commun 15:53057. https://doi.org/10.1038/s41467-024-53057-z

O'Brien E, Xagoraraki I (2019) A water-focused One-Health approach for early detection and prevention of viral outbreaks. One Health 7:100094. https://doi.org/10.1016/j.onehlt.2019.100094

Ocasio DV et al (2023) Cholera outbreak—Haiti, September 2022–January 2023. MMWR

of Science, O. (2009a). Method 1600: Enterococci in water by membrane filtration using membrane-Enterococcus Indoxyl-$-D-glucoside agar (mEI)

of Science, O. (2009b). Method 1603: Escherichia coli (E. coli) in water by membrane filtration using modified membrane-thermotolerant Escherichia coli agar (modified mTEC)

OMS (2022) Guidelines for drinking-water quality: fourth edition incorporating the first and second addenda. World Health Organ 33(33):1–614. https://iris.who.int/rest/bitstreams/907844/retrieve

Ong NTJ, Yee SK, Ong J, Ting N (2020) Review on water quality monitoring technologies. Indones J Electr Eng Comp Sci 18(3):1416–1423. https://doi.org/10.11591/ijeecs.v18.i3.pp1416-1423

Paerl HW et al (2011) Controlling harmful cyanobacterial blooms in a hyper-eutrophic lake (Lake Taihu, China): the need for a dual nutrient management strategy. Water Res 45(5):1973–1983

Pakistan Health Authorities (2018) Typhoid—Pakistan. WHO Disease Outbreak News

Park J-H, Bae K-S, Kang J, Yoon J-K, Lee S-H (2024) Comprehensive assessment of multidrug-resistant and extraintestinal pathogenic *Escherichia coli* in wastewater treatment plant effluents. Microorganisms 12(6):1119. https://doi.org/10.3390/microorganisms12061119

Paul MJ, Meyer JL (2001) Streams in the urban landscape. Annu Rev Ecol Syst 32:333–365. https://doi.org/10.1146/annurev.ecolsys.32.081501.114040

Per- and Polyfluoroalkyl Substances (PFAS) US EPA. Available from: https://www.epa.gov/sdwa/and-polyfluoroalkyl-substances-pfas? Accessed 17 Sept 2025

PFAS National Primary Drinking Water Regulation. 2024. https://www.epa.gov/system/files/documents/2024-04/pfas-npdwr_fact-sheet_general_4.9.24v1.pdf?utm

Piarroux R, Barrais R, Faucher B, Haus R, Piarroux M, Gaudart J, Magloire R, Raoult D (2011) Understanding the cholera epidemic, Haiti. Emerg Infect Dis 17(7):1161–1168. https://doi.org/10.3201/eid1707.110059

Pitt SJ, Gunn A (2024) The One Health concept. Br J Biomed Sci 81:12366. https://doi.org/10.3389/bjbs.2024.12366

Prata JC (2022) A One Health perspective on water contaminants. Water Emerg Contam Nanoplast 1:15. https://doi.org/10.20517/wecn.2022.14

Prüss-Ustün A, Wolf J, Bartram J, Clasen T, Cumming O, Freeman MC et al (2019) Burden of disease from inadequate water, sanitation and hygiene for selected adverse health outcomes: an updated analysis with a focus on low- and middle-income countries. Int J Hyg Environ Health 222(5):765–777. https://doi.org/10.1016/j.ijheh.2019.05.004

Popoola BM (2025) Antibiotic resistant bacteria in tropical freshwater ecosystems. Microbe 8:100457

Rockström J et al (2009) A safe operating space for humanity. Nature 461:472–475. https://doi.org/10.1038/461472a

Rose JB, Hofstra N, Hollmann E, Katsivelis P, Medema GJ, Verbyla ME (2023) Global microbial water quality data and predictive analytics: Key to Health and meeting SDG 6. PLOS Water 2(8):e0000166. https://doi.org/10.1371/journal.pwat.0000166

Rout AK, Rout SS, Panda A, Tripathy PS, Kumar N, Parida SN et al (2025) Potential applications and future prospects of metagenomics in aquatic ecosystems. Gene:149720

Rupa FH et al (2024) Addressing public health risks: strategies to combat waterborne diseases in Bangladesh. J Prev Med Public Health 24(3):495–505. https://pmc.ncbi.nlm.nih.gov/articles/PMC11626110/

Sambaza SS, Naicker N (2023) Contribution of wastewater to antimicrobial resistance: a review article. J Glob Antimicrob Resist 34:23–29. https://doi.org/10.1016/j.jgar.2023.05.010

Schwabe CW (1986) Veterinary medicine and human health. Kongressberaettelse-Nordiska Veterinaermoetet [Plenarfoeredrag], Sweden, p 15

Senta I, Rodríguez-Mozaz S, Corominas L, Petrovic M (2020) Wastewater-based epidemiology to assess human exposure to personal care and household products–a review of biomarkers, analytical methods, and applications. Trends Environ Anal Chem 28:e00103

Sims N, Kasprzyk-Hordern B (2020) Future perspectives of wastewater-based epidemiology: Monitoring infectious disease spread and resistance to the community level. Environ Int 139:105689

Sishu FK, Tilahun SA, Schmitter P, Assefa G, Steenhuis TS (2022) Pesticide contamination of surface and groundwater in an Ethiopian highlands' watershed. Water 14(21):3446. https://doi.org/10.3390/w14213446

Stage 1 and Stage 2 Disinfectants and Disinfection Byproducts Rules. US EPA. (n.d.). Available from: https://www.epa.gov/dwreginfo/stage-1-and-stage-2-disinfectants-and-disinfection-byproducts-rules. Accessed 16 Sept 2025

Su Y, Zhang Z, Zhu J et al (2021) Microplastics act as vectors for antibiotic resistance genes in landfill leachate: the enhanced roles of the long-term aging process. Environ Pollut 270:116278. https://doi.org/10.1016/j.envpol.2020.116278

Sugimoto JD, Koepke AA, Kenah EE, Halloran ME, Chowdhury F, Khan AI, Longini IM (2014) Household transmission of Vibrio cholerae in Bangladesh. PLoS Negl Trop Dis 8(11):e3314. https://doi.org/10.1371/journal.pntd.0003314

Suttner B (2020) Metagenomics approaches for improved water quality monitoring, microbial source tracking, and public health risk assessment. Doctoral dissertation. Georgia Institute of Technology, Atlanta, GA

The Guardian (2024) 'Rivers you think are pristine are not': how drug pollution flooded the UK's waterways. The Guardian. https://www.theguardian.com/environment/2024/sep/27/amr-drug-resistance-england-national-parks-hidden-hazards-rivers-pollution-aoe

The Guardian (2025) Sudan cholera outbreak kills 40 in a week as health centres overwhelmed. The Guardian

Times of India (2017) Jaipur hospitals witness rise in diarrhoea. The Times of India

Times of India (2018) Jaipur hospitals see surge in diarrhoea cases. The Times of India

UK Health Security Agency (UKHSA) (2025) Travel-associated enteric fever cases in 2024. UKHSA. GOV.UKVax-Before-Travel

US EPA Final PFAS NPDWR (pre-publication Federal Register notice) - Hazard Index for mixtures incl. PFBS. 2025. US EPA

Vörösmarty CJ et al (2010) Global threats to human water security and river biodiversity. Nature 467:555–561. https://doi.org/10.1038/nature09440

WHO. Nitrate and Nitrite in Drinking-water Background document for development of WHO Guidelines for Drinking-water Quality. Available from: http://www.who.int/publications/guidelines/. Accessed 16 Sept 2025

World Health Organization (WHO) (2017) Guidelines for drinking-water quality (4th ed., incorporating the 1st addendum). World Health Organization. https://www.who.int/publications/i/item/9789241549950

World Health Organization (WHO) (2022) One Health. WHO. Available from https://www.who.int/news-room/fact-sheets/detail/one-health

World Health Organization (2024) Guidelines for drinking-water quality: small water supplies. World Health Organization

World Health Organization. WHO small-supplies executive summary (companion overview). 2025. Accessed 16 Sept 2025

Yang Y, Liu G, Song W et al (2019) Plastics in the marine environment are reservoirs for antibiotic and metal resistance genes. Environ Int 123:79–86. https://doi.org/10.1016/j.envint.2018.11.061

Zhu Y-G, Johnson TA, Su J-Q et al (2013) Diverse and abundant antibiotic resistance genes in Chinese swine farms. PNAS 110(9):3435–3440. https://doi.org/10.1073/pnas.1222743110

Zhu C et al (2025) Global diversity and distribution of antibiotic resistance genes. Nat Commun

MIX
Papier | Fördert gute Waldnutzung
FSC® C083411

Zeitfracht Medien GmbH
Ferdinand-Jühlke-Straße 7
99095 Erfurt, Deutschland
produktsicherheit@kolibri360.de